ORBITALS, TERMS and STATES

ORBITALS, TERMS and STATES

Malcolm Gerloch
University Chemical Laboratory, Lensfield Road, Cambridge

JOHN WILEY & SONS
Chichester · New York · Brisbane · Toronto · Singapore

Library of Congress Cataloging-in-Publication Data:

Gerloch, M.
Orbitals, terms and states.

Includes index.
1. Molecular orbitals. I. Title.
QD461.G37 1986 541.2′2 85–26432

ISBN 0 471 90935 1 (cloth)
ISBN 0 471 90936 X (paper)

British Library Cataloguing in Publication Data:

Gerloch, Malcolm
Orbitals, terms and states.
1. Quantum chemistry
I. Title
541.2′8 QD462

ISBN 0 471 90935 1 (cloth)
ISBN 0 471 90936 X (paper)

Printed and bound in Great Britain

for
Linda & Pinki
who played
their part

CONTENTS

PREFACE

I believe that many chemistry students are unclear about the nature of orbitals, terms, levels and states; and that they feel a need for a small book about them. It is all very well to say that the material has already been published many times in various standard texts—and it has, and some are splendid—but it is a fact that those sources are either dated, slanted too much towards physics, or too comprehensive. If the latter, the area is not unfolded as much as clear pedagogy might require and is, in any case, lost within the daunting length of a large work. So I have presented here a selection from atomic and molecular spectroscopy, from quantum mechanics and bonding theory, that I hope forms a satisfying nucleus from which understanding of those subjects can grow. In addition to the title topics, I offer brief introductions to the expansion theorem and the concept of basis functions; perturbation theory and the variational principle; secular equations and determinants; Slater determinants and antisymmetry; real versus complex orbitals; quantum numbers and symmetry labels. Explanations of, and simple manipulations with, the angular momentum operators L^2, L_z, S^2, S_z, J^2, J_z are provided within the contexts of both free-ions and linear molecules. And central to full discussions of angular momentum presented here, are the meanings of symbols like:

$s, p, d, f \ldots$	$p^2, d^3 \ldots$
$S, P, D, F \ldots$	${}^2D, {}^4F, {}^3F_4, {}^6S_{5/2} \ldots$
$l, s, j, m_l, m_s, m_j \ldots$	${}^1\Sigma_g^+, {}^3\Delta_u.$
$L, S, J, M_L, M_S, M_J \ldots$	

I have attempted to minimize the mathematical content as far as possible but commensurate with reality. However, I have never believed that the eradication of all algebraic equations from chemistry texts leads to clarity, even for those who are unhappy with mathematics. The verbal complexity that comes with the avoidance of all equations almost invariably leaves behind obscurity; and, frequently, totally incorrect understanding. It is one thing to ask that the subject be made simple; quite another to be content with versions that are just simple-minded. Chemistry is one of the 'hard sciences' but that, at least, provides a source of pride as we master each new stage. Even so, I have presented virtually no full proofs in the text, adopting instead a sort of 'black box' approach. Rather full explanations are given for the 'input'—the physical nature of the problem to be discussed; and of the 'output'—the physical significance of mathematical results which can be derived. Provided that the middle bit involves no further physical input and is 'merely' mathematical manoeuvre, we regard it as a 'black box' in this book and rarely discuss it.

The dauntless questions of several generations of tutorial students have done much to shape the presentation of my material and I thank them. I suppose it is too much to expect that all similar questions from their successors will now be unnecessary. Again, it is my pleasure to thank Professor Freeman and his colleagues in the University of Sydney for their hospitality while much of this book was being written; my research students who put up with endless 'readings'; and finally Gwyneth Neal-Freeman for her transcription of my none-too-tidy manuscript.

MG
Cambridge, 1985

CHAPTER 1

Introduction

1.1 *Leaping to a start*

In addressing students reading for an Honours degree in Chemistry, I share with other authors a duty to limit my scope to what is desirable and pedagogically achievable. I *know* that many students experience difficulty with Russell–Saunders coupling, with orbitals, terms and states, and with bases: and so I believe that this book will be helpful. My dilemma, however, is how to reach a suitable starting point without first writing another book on the principles of theoretical chemistry. As for a bather by a pool, I see only one way: it is to jump. It requires cheek on my part and nerve on the student's. And there is no reason why we should not have some fun on the way.

Classical mechanics is about Newton's three laws and their consequences. The first says that bodies move with uniform velocities unless acted upon by a force. Very sensible—obvious, maybe: and the third law about action and reaction is demonstrable simply by banging your head against a wall—it hurts every time. It is only a little more difficult to verify the second law. Actually, we do not have to express classical mechanics in this way. Instead we can invoke the laws of conservation of energy and momentum and then Newton's first two laws can be derived from these. Which are more fundamental—Newton's laws or those of conservation? No *objective* choice can be made. All we have are three distinct principles, that may be expressed in a variety of ways, and upon which rests the edifice we call classical mechanics. *Why* does a body continue in a straight line unless acted upon by a force? Why does $p = ma$? Science provides no answer. Science classifies and correlates phenomena: we are in the business of constructing circles.

Well, as we all know, there came a time when classical mechanics failed to account for experiment—the ultraviolet catastrophe and so on. Quantum mechanics was born. Actually the birth was a two-stage affair and many texts written earlier this century, and to which we all turn, devote an apparently unseemly amount of space to the transition from classical to the 'old quantum theory' and then to the 'new quantum theory'. *We* shall be brave and leap right through. What are the axioms of quantum mechanics? Well, for one thing, they are not so readily describable as those for the classical world. They are, however, equally non-fundamental in the sense that different sets of axioms may be chosen: once again, all that matters is a self-consistent and complete quantum edifice. Secondly, note that all the various axioms must usually be stated, or perhaps partly stated, mathematically. It is frequently the case that chemistry students are a little scared of mathematics and it is sometimes held that therein lies their unease with quantum mechanics itself. That could be so but it may be of some psychological assistance here to observe that the fundamentals of quantum mechanics really *are* difficult to understand (though not actually because of the mathematical context) and, indeed, oblige one to think again about what the word 'understanding' really means.

It is worth our while thinking briefly about the psychology of learning this subject. The problem is that notions like 'an electron is both a wave and a particle' or 'certain quantities cannot be assigned definite values' simply do not square with common sense. The correct response, however, is not to doubt one's common sense. Common sense means common experience and we do not have any common experience of microscopic objects like electrons or protons. Why on earth should we *intuitively* require fundamental particles to obey classical, or indeed any, particular mechanics? Providing they obey Newtonian mechanics when aggregated into macroscopic objects like a half-brick, that will be fine. This remark is, of course, a (cavalier) way of stating the *correspondence principle*. Now consider how we might try to test the classical equation, $s = ut + \frac{1}{2}at^2$, but for the motions of fundamental particles. Well, first think how we would do it for macroscopic objects. We might let a half-brick fall under gravity and time its progress as it passes a number of points. Perhaps we would arrange that it catches some electronic microswitches as it falls: maybe not, though, for these may interfere with its free fall sufficiently to invalidate our experiment. Better use some form of optical switching and let the brick cut through light beams. Nobody with common sense needs persuading that we can arrange to test, and indeed confirm, that classical equation with as much accuracy as patience and money deem worthwhile. But return to the situation with a fundamental particle. In order to get an accurate answer, we need to observe its motion without disturbing its progress. We must choose some apparatus more delicate, less energetic than the object we observe. But what is more delicate, more fundamental than a fundamental particle? Scaling up to the macroscopic world again: it would be like† monitoring the motion of a half-brick by throwing other half-bricks at it and analysing their deflected trajectories. Much joy to the experimenter!

All this is loosely about the *uncertainty principle*. The philosophical core of the sketch, however, is to emphasize the fact that we must now view science as relating *experiments* rather than as providing a description of the world as it 'really is'. Given a familiarity with the uncertainty principle, we must resist asking questions like, 'I know we cannot measure x and p_x simultaneously and exactly, but, nevertheless what is the electron *actually* doing? The scientific answer has to be 'We don't know and, as far as current quantum theory is concerned, we cannot know: the question belongs in metaphysics, not here'. When a student complains to me that he does not 'understand' quantum mechanics, I can only reply 'What makes you so special—neither do I'. Unless, that is, we can come to terms with what is meant by that word 'understand'. After all, do we understand *classical* mechanics? We cannot provide the answer to *why* Newton's laws are obeyed. All attempts to do so are merely illustrations of the fact that they are; parts of a grand circle. Maybe we should say that, 'Understanding is a state of mind achieved when you cease to question'. Accept Newton's three laws and all will be well—on the classical scale anyway. Accept the axioms of quantum mechanics and all will be well on the microscopic scale also. Do not ask how an electron can be both a wave and a particle. We cannot sense it anyway so suspend common sense. Use the equations and rules of quantum mechanics—by rote if you like—until you are familiar with them. In due course you will forget to ask those awkward—but human—ques-

† This is very specious logic: I leave it in because I wish to sketch a feeling very quickly. For clarity and accuracy but at some greater length, see Polkinghorne,[1] for example.

tions and that Zen-like state of understanding will have been achieved.

1.2 *Some basic hardware*

The sort of algebra we use in quantum mechanics has to be different from that employed in the classical arena. Consider the Newtonian expression $p=ma$ and note two things. First, given definite values for m and a, we inevitably calculate a definite value for p—the algebra obliges it—so there are bound to be occasions in quantum mechanics where another type of equation is necessary, structured so as to admit the notion of uncertainty. Secondly, the symbol p, for example, stands for two things—the (unspecified) numerical value of the momentum, and the abstract quality of momentum. While this distinction is pedantic in the extreme in the classical area, a separation of sorts is quite evident in the quantum field.

A central equation used in much of quantum mechanics is the so-called *eigenvalue equation*. It has the form:

$$\text{(operator) on (function)} = \text{(scalar)} \times \text{(same function)}$$
$$\hat{O}(f) = af \tag{1.1}$$

A function that has this special property with respect to the operator is called an *eigenfunction* (proper, or characteristic, function) of $\hat{O}$; and its accompanying scalar a is called an *eigenvalue*. The operator is to be associated with the physical quantity of immediate interest and the eigenvalue with the numerical value that experiment will yield for that quantity: here is the 'separation' of quality and quantity, referred to above. Suppose we wish to measure some property B of a particular system, we proceed as follows:

(a) Construct an operator $\hat{B}$ that represents that property in that system.
(b) Discover some function that satisfies the equation $\hat{B}f = bf$, and
(c) Identify b with that numerical value which may be observed for the property in question.

A most common example—for us—is to find the total energy of some given system. The energy operator has a special name, deriving from a generalized form of classical mechanics. It is called the Hamilton operator and is often given the symbol $\mathscr{H}$. Equally conventionally, eigenfunctions of H are usually written with Greek symbols ψ, ϕ or χ: none of these are mandatory of course. So the eigenvalue equation relevant for a study of energy is:

$$H\psi = E\psi \tag{1.2}$$

and the eigenvalue E gives that value of the energy, associated with the function ψ, and which is experimentally and exactly observable. An interpretation of the eigen*function*—which, for historical reasons connected with the 'old' quantum theory, is also called a wavefunction—is to say that it describes a situation or state in which the constituent particles of the system in question find themselves. Those particles may assume some other state ψ' that is also an eigenfunction of H,

$$H\psi' = E'\psi' \tag{1.3}$$

in which case we observe *discrete* eigenvalues—here energies—for the system. This mathematical structure thus naturally throws up the central notion of quantum mechanics—that observables are usually quantized.

In order to exploit the procedure outlined above we need to know how to construct the operators of interest. Essentially the 'rules of the game' are as follows:

(i) Formulate a classical description of the property of interest in the given system at some instant in time,
(ii) Replace the classical quantities in that description with operators, according to the fixed prescription below, and
(iii) The eigensolutions of the operator so formed will yield time-averaged descriptions for the quantum-mechanical system.

These statements are a bit 'rough-and-ready' and require some clarification. First, however, we state the recipes for operator formation. In this book we only discuss stationary states and time-independent situations. It is enough then to provide the replacements:

	Classical quantity	→ Quantum operator
Position:	$x, y, z, \mathbf{r}$	$\rightarrow \hat{x}, \hat{y}, \hat{z}, \hat{\mathbf{r}}$

Momentum: $p_x, p_y, p_z, \mathbf{p} \rightarrow -i\hbar\frac{\partial}{\partial x}, -i\hbar\frac{\partial}{\partial y}, -i\hbar\frac{\partial}{\partial z}, \hat{\mathbf{p}}$

We place carets on operators when ambiguity might otherwise arise. The operator $\hat{x}$ means 'multiply what follows by x': e.g. $\hat{x}y = xy$. The replacement of momentum components p_α by $-i\hbar\,(\partial/\partial\alpha)$ (where $\hbar = h/2\pi$) often causes 'conceptual' problems. So it ought. After all, one cannot conceive of a common-sense meaning for it and I personally gravely doubt that an 'explanation' for it in terms of de Broglie waves does more than shift one's sense of unreality from one place to another. It is far simpler and more practical to accept this *seemingly* arbitrary replacement simply as a recipe. Once more, asking 'why' about axioms is rather unrewarding. As things are developed, it will become more obvious how this route leads to desired—that is, observed—results. However, as we shall not discuss all aspects of the quantum chemistry to be found in a degree course, let us note now for completeness' sake, that the imaginary i is put there to help form an hermitian operator and hence ensure that the ensuing eigenvalues are *real*, as they must be to be observable. Other points arising from the given rules are best made by example.

Let us construct a quantum description for the energy of a freely moving particle of mass m. Application of rule (i) leads to the classical description of its energy—here kinetic only—as $\frac{1}{2}mv^2$. Suppose we choose to align a reference frame so that the motion takes place parallel to x. The (kinetic) energy is then $\frac{1}{2}mv_x^2$ or, since $p_x = mv_x$, $p_x^2/2m$. Rule (ii) requires the replacement of p_x by $\hat{p}_x$, leading to the energy operator,

$$H = -\frac{\hbar^2}{2m}\frac{\partial^2}{\partial x^2} \cdot \tag{1.4}$$

In step (iii) we can find a solution from memory: what functions when double-differentiated yield themselves? We can take sines, cosines or their complex combinations, exponentials. For example, with sine functions,

$$-\frac{\hbar^2}{2m}\frac{\partial^2}{\partial x^2}(\sin nx) = \frac{n^2\hbar^2}{2m}\sin nx. \tag{1.5}$$

The constant n does not take any value but only positive, integral ones, as we discuss in Chapter 6. Even so we clearly have an infinitely large set of eigenfunctions, $\{\sin nx\}$, each with its discrete, quantized, eigenvalue—or energy in this case. Had we chosen an arbitrary reference frame in the first place, the classical expression for the kinetic energy would be written as

$$\tfrac{1}{2}mv^2 = \frac{1}{2m}(p_x^2 + p_y^2 + p_z^2), \tag{1.6}$$

and the quantum energy operator H as,

$$H = -\frac{\hbar^2}{2m}\left(\frac{\partial^2}{\partial x^2} + \frac{\partial^2}{\partial y^2} + \frac{\partial^2}{\partial z^2}\right). \tag{1.7}$$

The quantity in brackets occurs frequently and is written

$$\left(\frac{\partial^2}{\partial x^2} + \frac{\partial^2}{\partial y^2} + \frac{\partial^2}{\partial z^2}\right) \equiv \nabla^2, \tag{1.8}$$

pronounced 'del-squared'. So the Hamiltonian operator for a particle possessing only kinetic energy is most generally written as

$$H = -\frac{\hbar^2}{2m}\nabla^2 \tag{1.9}$$

Its eigenfunctions must now be functions of all three spatial coordinates and so are more complex than the $\sin nx$ function described above.

1.3 *Atomic Hamiltonians*

We move on quickly now to a problem of much greater interest and chemical significance. In so doing, further clarification of the 'game rules' can be made. Consider the hydrogen atom. In a so-called fixed-nucleus approximation, we envisage that atom as a stationary proton associated with which is a captive, moving electron. We shall construct the Hamiltonian for the electron of mass m and charge $-e$. Classically, at a frozen instant in time, the electron is moving in some general direction with velocity v and so has kinetic energy T, given by (1.5). At that same instant, it lies at distance r

from the proton of charge e, and possess potential energy V given by,

$$V = -\frac{e^2}{r}, \tag{1.10}$$

the minus sign arising from the convention that the potential energy of a bound electron decreases to a minimum as it is brought *in* from an infinite distance at an energy zero. Making the usual operator substitutions then yields the electronic Hamiltonian for the hydrogen atom as

$$H = -\frac{\hbar^2}{2m}\nabla^2 - \frac{e^2}{r}. \tag{1.11}$$

Perhaps here is the place to emphasize that there is no such thing as *the* Hamiltonian. It is an energy operator tailored to the problem at hand. So H in (1.9) for a free particle differs from H in (1.11) for the hydrogen electron, and so on. As the energy operator changes, so also do the forms of solutions to the energy eigenvalue equation. Eigenfunctions, ϕ, of H_H which we shall write for that in (1.11) are solutions for what is also called the Schrödinger equation for the hydrogen electron,

$$\left(-\frac{\hbar^2}{2m}\nabla^2 - \frac{e^2}{r}\right)\phi = \epsilon\phi, \tag{1.12}$$

and are functions of three-dimensional space and again more complicated than those for either (1.4) or (1.9). The reader will know that we call these functions 'orbitals' and is familiar with their descriptions by labels like 1*s*, 2*s*, 2*p*, . . . 3*d* . . . 4*f* . . . It is part of the subject-matter of this book to explain those labels rather fully. Before leaving the hydrogen atom in this brief introduction, recall that physical interpretations of wavefunctions and orbitals like these ϕ are made in terms of the quantities ϕ^2—or if they are complex, $\phi^*\phi$—as describing probability distributions; in this case, electron densities. This is what was intended—and all that was intended—by the caveat about 'time-averaged descriptions' in rule (iii) above. The point is that the construction of the Hamiltonian began with a consideration of the 'time-frozen' classical situation and in due course this leads to the statistical distribution inherent in our interpretation of the wavefunction ϕ.

Let us increase the classical complexity of our example slightly. Consider, instead of the hydrogen atom, a hydrogen-like atom. By this is meant an atom in which a single electron may be viewed as moving essentially outside closed core shells. For example, the ions Ti^{3+} with the configuration $Ar3d^1$. A Hamiltonian operator designed to describe the behaviour of any one 'outer' electron would be very similar to that in (1.11). All we need to change is the numerator in the potential energy operator in recognition that the outer electron is now subject to a nuclear charge that is screened by the core: we refer to the effective nuclear charge $Z_{\text{eff}}e$ and so write

$$\mathcal{H}_{\text{H-like}} = -\frac{\hbar^2}{2m}\nabla^2 - \frac{Z_{\text{eff}}e^2}{r} \tag{1.13}$$

It is hopefully easy to accept that solutions for this Hamiltonian—also called orbitals—are very similar to those for hydrogen but spatially scaled, though not exactly linearly. They are still labelled by the letters *s*, *p*, *d* . . .

Compare, however, the situation for a so-called 'many-electron' atom. Here we refer to more than one electron outside of closed shells; for example, V^{3+} ions with configuration $Ar3d^2$. The instantaneous situation to which we apply rule (i) is represented in Figure 1.1. The kinetic energy of electron number 1 is $\frac{1}{2}mv_1^2$ and of electron 2, $\frac{1}{2}mv_2^2$. The kinetic energy Hamiltonian operator is then

$$\mathcal{H}_{\text{KE}} = \frac{-\hbar^2}{2m}\nabla_1^2 - \frac{\hbar^2}{2m}\nabla_2^2 \tag{1.14}$$

where the subscripts indicate that each operator acts upon the coordinates of only one electron. We call these 'one-electron operators' because of that. In the more general case of *n* electrons outside the core, the kinetic energy Hamiltonian is given by the sum

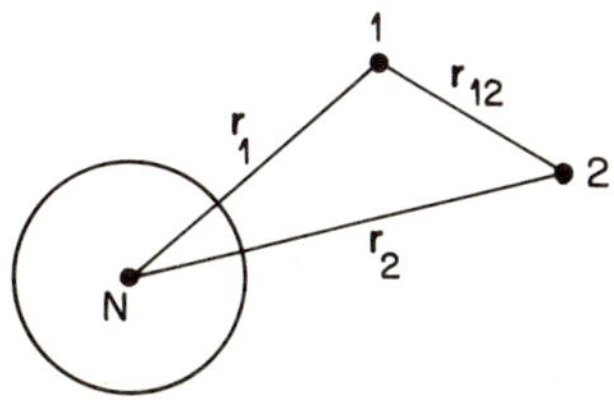

Figure 1.1 Two electrons outside closed shells.

$$\mathcal{H}_{\mathrm{KE}} = -\frac{\hbar^2}{2m}\sum_i^n \nabla_i^2 \,, \qquad (1.15)$$

where m is taken outside the summation sign only because we refer here to similar particles, of course. The classical potential energy of the electrons in Figure 1.1 is the sum of parts describing the individual (discrete) attraction of each electron by the shielded nucleus and described by terms like that in (1.13), with an additional part representing the mutual electron repulsion:

$$\mathcal{H}_{\mathrm{PE}} = -\frac{Z_{\mathrm{eff}}e^2}{r_1} - \frac{Z_{\mathrm{eff}}e^2}{r_2} + \frac{e^2}{r_{12}} \qquad (1.16)$$

Again, generalizing to an n-electron atom:

$$\mathcal{H}_{\mathrm{PE}} = -Z_{\mathrm{eff}}\sum_i^n \frac{e^2}{r_i} + \sum_{i<j}^m \frac{e^2}{r_{ij}} \,. \qquad (1.17)$$

The restriction on i in the second summation prevents any double counting of interelectron repulsions. Putting (1.15) and (1.17) together with (1.13) we get our complete electronic Hamiltonian for the outer electrons of n-electron atoms as,

$$\mathcal{H}_{\mathrm{many}} = \sum_i^n \mathcal{H}_{\text{H-like}}(i) + \sum_{i<j}^n \frac{e^2}{r_{ij}}. \qquad (1.18)$$

The 'interelectronic repulsion operator' (not a good name, though, as we see in Chapter 5) is called the Coulomb operator.

Now this Hamiltonian has a much more demanding structure than that for hydrogen-like atoms—because of the Coulomb operator. There is no reason at all at this stage to expect that the eigensolutions ψ of

$$\mathcal{H}_{\mathrm{many}}\,\psi = E\psi \qquad (1.19)$$

should resemble the orbitals ϕ of

$$\mathcal{H}_{\text{H-like}}\,\phi = \epsilon\phi \,. \qquad (1.20)$$

For one thing, the Coulomb operator is a 'two-electron operator'—because it describes, through $1/r_{12}$, electronic *inter*actions—and so the mechanics of the electrons are inextricably interlinked or *coupled*. So we have a great difficulty here. Despite this, chemistry students' first acquaintance with quantum numbers and the like, includes the aufbau principle and an apparent implication that one-electron orbitals with labels like *s*, *p*, *d* . . . are a uniform property of all atoms. Is that valid or not? Well, this book is much concerned with just this issue and we have now sketched in enough common ground for our study of orbitals, terms and states to begin.

CHAPTER 2

Angular Momentum

2.1 *Angular momentum operators*

Since the avowed intention of this book is to explain orbitals and states in a way that is transparent and unfrightening, it might appear perverse to begin with a topic so seemingly bereft of chemical relevance as angular momentum: certainly it will be emphasized repeatedly throughout *this* chapter that initially we need not have electrons or atoms in mind at all and that we focus deliberately on the angular momentum properties of *any* object whatever. Yet a little generality and abstraction here pays off handsomely in all that follows.

Consider, then, any rigid or point object moving along an arbitrary trajectory in three-dimensional space. Such motion can always be decomposed at any instant into a sum of translational and rotational components. In Figure 2.1(b), the instantaneous rotational motion of an object distant r from the instantaneous centre of

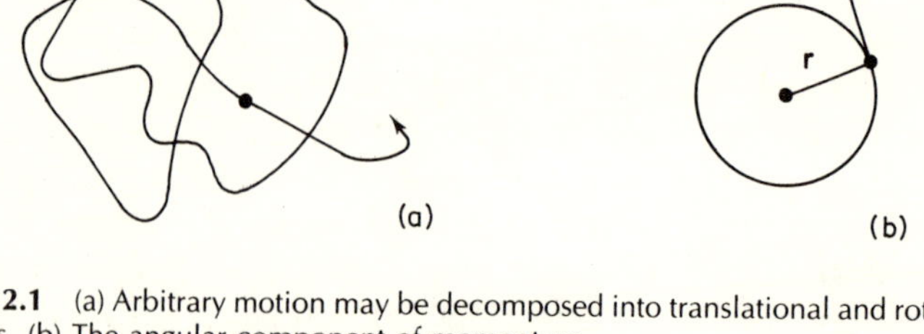

Figure 2.1 (a) Arbitrary motion may be decomposed into translational and rotational parts. (b) The angular component of momentum.

rotation is described by an angular momentum **l** about that centre. If the instantaneous linear momentum is **p** then

$$\mathbf{l} = \mathbf{r} \wedge \mathbf{p}, \tag{2.1}$$

where all these quantities have direction and magnitude and so are vectors (printed **bold**). So despite the generality of the arbitrary complex motion illustrated in Figure 2.1(a), the instantaneous angular momentum of any object can be studied through an investigation of the simple situation shown in Figure 2.1(b) and by equation (2.1).

With the help of Figure 2.2 we can write down the components of the vector **l** with respect to some arbitrarily fixed cartesian reference frame. We take the cartesian axis **z** as directed up, out of

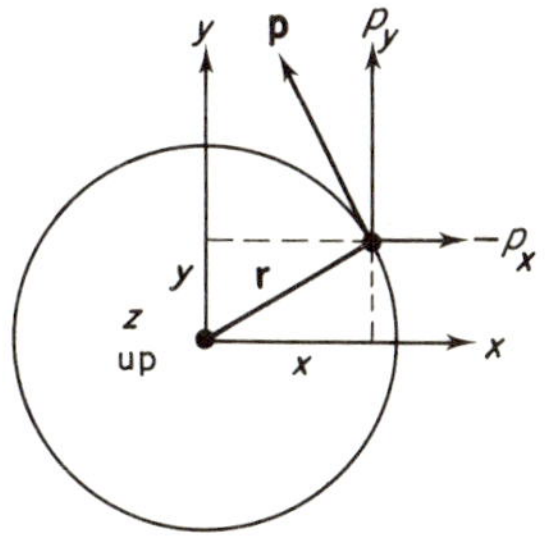

Figure 2.2 Resolving components of the vectors **r** and **p** in a right-handed frame.

the page, by convention so as to define a right-handed coordinate frame: sometimes the handedness chosen for reference frames is of little concern but often the signs of various quantities *are* sense-dependent. To maintain maximal correspondence with the literature, it is advisable to use right-handed frames exclusively. We define positive angular momentum with respect to any one axis as involving movement of the particle from one axis towards that later in the alphabetical sequence. Thus for the component of angular momentum about the z-axis, as in Figure 2.2, we examine the situation of anticlockwise rotation—that is, from x towards y—as viewed from the positive z direction towards the origin. Components with respect to the x and y axes are defined by cyclic permutations of the axis labels. So, while the components of **r** in the

xy-plane are just x and y, those of **p** are $+p_y$ but $-p_x$, as shown. Hence the angular momentum with respect to z is given as

$$l_z = x(p_y) + y(-p_x)$$

or

$$l_z = xp_y - yp_x. \qquad (2.2)$$

Similar constructions can be made for the yz and zx planes, but corresponding expressions for l_x and l_y are obtained more immediately by cyclic permutations of axis labels in (2.2); once more, in recognition of the isotropy of free space:

$$l_x = yp_z - zp_y \qquad (2.3)$$

$$l_y = zp_x - xp_z\ . \qquad (2.4)$$

Now we may use the operator correspondences described in Chapter 1 to establish the equivalent quantum mechanical *operators* for angular momentum. Replacing the coordinate x by the operator $\hat{x}$, and the momentum p_x by $-i\hbar(\partial/\partial x)$, etc., we arrive at the angular momentum operators

$$\hat{l}_z = -i\hbar\left(x\frac{\partial}{\partial y} - y\frac{\partial}{\partial x}\right) \qquad (2.5)$$

$$\hat{l}_x = -i\hbar\left(y\frac{\partial}{\partial z} - z\frac{\partial}{\partial y}\right) \qquad (2.6)$$

$$\hat{l}_y = -i\hbar\left(z\frac{\partial}{\partial x} - x\frac{\partial}{\partial z}\right). \qquad (2.7)$$

The operator $\hat{\mathbf{l}}$ for the total orbital angular momentum of an arbitrary object comprises the vectorial sum of these components; namely

$$\mathbf{l} = \mathbf{i}l_x + \mathbf{j}l_y + \mathbf{k}l_z. \qquad (2.8)$$

We have more need, however, of the square of this quantity, $l^2 = \mathbf{l}.\mathbf{l}$,

$$l^2 = l_x^2 + l_y^2 + l_z^2, \tag{2.9}$$

which because of the nature of the scalar, or dot product, contains no cross-terms, of course.

Let us review the position reached so far. We considered an arbitrary motion of an arbitrary object and established the classical descriptions (2.2)–(2.4) for the components of the angular momentum constituent of the motion at any given instant in time. Use of the basic correspondences of quantum mechanics then allowed us to write down corresponding quantum mechanical operators for these quantities, (2.5)–(2.7), and for the square of the total angular momentum (2.9). All these still refer to the angular momentum of an arbitrary object.

2.2 *The commutation rules*

If the last section was physics formalized by some mathematics, what follows begins with pure algebra spiked with a little engineering. In this spirit, let us play.

The commutator of two quantities (here, operators), a and b, is written as a so-called Lie bracket and defined by

$$[a, b] \equiv ab - ba. \tag{2.10}$$

If $[a, b] = 0$, the order of multiplication of a and b is unimportant and the operators a and b are said to commute.

Before going further, consider very practical circumstances which illustrate the worldly relevance of non-commutative algebra. Let a and b represent the operations of rotation by 90°—say clockwise—as seen looking down each of two orthogonal axes towards the origin. We apply these operations upon an object in different orders, as illustrated in Figure 2.3, and get very different answers R_1 and R_2. The engineer who determines the sequence of drilling and rotating during the manufacture of a car engine cylinder block had better get it right or the consequences of non-commutative algebra will be stark indeed! So we might hope for some worthwhile results to emerge from a study of the commutation properties of angular momentum operators: they turn out, in fact, to be amazingly far-reaching.

The beginning of the process is a determination of the commutator of $\hat{l}_x$ and $\hat{l}_y$:

$$
\begin{aligned}
[l_x, l_y] &\equiv l_x l_y - l_y l_x \\
&= \left\{-i\hbar\left(y\frac{\partial}{\partial z} - z\frac{\partial}{\partial y}\right).-i\hbar\left(z\frac{\partial}{\partial x} - x\frac{\partial}{\partial z}\right)\right\} \\
&\quad - \left\{-i\hbar\left(z\frac{\partial}{\partial x} - x\frac{\partial}{\partial z}\right).-i\hbar\left(y\frac{\partial}{\partial z} - z\frac{\partial}{\partial y}\right)\right\} \\
&= -\hbar^2\left\{\left(y\frac{\partial}{\partial x} + yz\frac{\partial^2}{\partial z\partial x} - yx\frac{\partial^2}{\partial z^2} - z^2\frac{\partial^2}{\partial y\partial x} + zx\frac{\partial^2}{\partial y\partial z}\right)\right. \\
&\quad \left. - \left(zy\frac{\partial^2}{\partial x\partial z} - z^2\frac{\partial^2}{\partial x\partial y} - xy\frac{\partial^2}{\partial z^2} + x\frac{\partial}{\partial y} + xz\frac{\partial^2}{\partial z\partial y}\right)\right\} \\
&= -\hbar\left(y\frac{\partial}{\partial x} - x\frac{\partial}{\partial y}\right) \\
&= i\hbar lz.
\end{aligned}
\tag{2.11}
$$

Thus l_x and l_y do not commute and the exact form of their commutator is important in the mathematical side of the theory we summarize later. Commutators of the pairs of angular momentum components are obtained by cyclic permutation once again. It can also be shown, albeit after some tedious algebra which we shall not bother with, that l^2 commutes with each of its components. Altogether there arises the set of *commutation relations* characteristic of angular momentum operators:

$$
\begin{array}{llllll}
[l_x, l_y] & = i\hbar l_z, & [l_y, l_z] & = i\hbar l_x, & [l_z, l_x] & = i\hbar l_y \\
[l^2, l_x] & = 0, & [l^2, l_y] & = 0, & [l^2, l_z] & = 0.
\end{array}
\tag{2.12}
$$

Once more, remember that these relationships are a property of angular momentum rather than of the particular nature of the body possessing that property.

2.3 *Observables*

At this point we require an operator theorem, which we state without proof. It is: *If two operators commute, they share a common set of eigenfunctions*. In symbols,

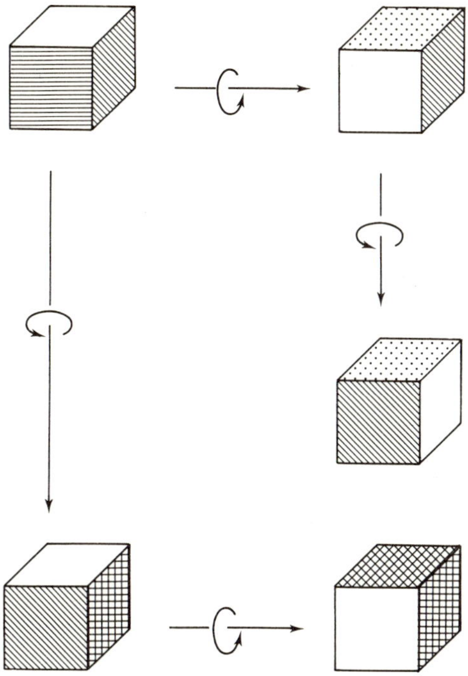

Figure 2.3 The order of two 90° rotations on this object is important: their non-commutation gives rise to two different results R_1 and R_2.

$$\text{if } [A, B] = 0, \qquad \begin{aligned} A\psi &= a\psi \\ B\psi &= b\psi. \end{aligned} \tag{2.13}$$

For example, consider the operators $\partial/\partial x$ and $\partial^2/\partial x^2$ which obviously commute. We know immediately that the set of functions $\{e^{nx}\}$ satisfy eigenvalue equations of both operators:

$$\frac{\partial}{\partial x} e^{nx} = n\, e^{nx}$$

$$\frac{\partial^2}{\partial x^2} e^{nx} = n^2 e^{nx}$$

However, the theorem above refers to the *sharing* of a common set of eigenfunctions: it does not require that *all* eigenfunctions of one operator also satisfy the other. In the present example observe that the set $\{\sin nx\}$ satisfies the second differential operator but not the first. Finally, note also that the converse theorem can be proved, namely: *two operators that share a common set of eigenfunctions, commute*.

These theorems lie wholly within the realm of mathematics, for no physical input is required in their proof. In keeping with the general policy of this book I have not taken the trouble, therefore, to present their derivations. However, as an 'aside' from our theme of angular momentum, it is worth illustrating the relevance of the theorems in a broader quantum mechanical context. Thus we write the operators relating to the position and momentum of a particle in the x dimension, say, as $\hat{x}$ and $\hat{p}_x$; that is as $\hat{x}$ and $-i\hbar(\partial/\partial x)$. These two operators clearly do not commute, for the action of differentiating a function of $f(x)$ and subsequently multiplying it by x gives a different result from one of multiplication by x first. Hence, according to our theorems, the position and momentum operators cannot share a common set of eigenfunctions. The physical significance of this important result emerges in quantum mechanics when we combine it with the postulated physical interpretation of an eigenfunction. Here we say that a *measurement* of the x-coordinate of a particle manifests the particle in an eigenstate—say ϕ—of the associated operator $\hat{x}$. Measuring the value of some other property whose associated operator $\hat{O}$ commutes with $\hat{x}$ would leave the particle in that same state and hence yield a unique eigenvalue of the quantity in question. On the other hand, a measurement of the momentum, associated with a non-commuting operator means that only a value of uncertain reliability is to be found, as we cannot know which eigenstate of $\hat{p}_x$ the system would take up at the instant that it is already fixed in state ϕ by the simultaneous measurement of the x-coordinate. In short, the simultaneous measurement of two properties associated with non-commuting operators cannot yield definite values for both: and herein lies the essence of Heisenberg's Uncertainty Principle. Conversely, definite values of quantities represented by *commuting* operators *can* be measured simultaneously. An example of just this is central to our development of the theory of angular momentum.

Looking back at the angular momentum commutation relationships in (2.12), we note that just two operators in the set $\{l^2, l_x, l_y, l_z\}$ commute—l^2 and any one of the components. Hence definite values of the total angular momentum squared and of the component parallel to any one axis can be determined and described simultaneously: we cannot measure components of angular momentum along more than one direction at the same time. It is a matter of *convention* that we chose a set of commuting angular momentum operators as l^2 and l_z and hence refer to z as the 'axis of quantization'.

Applying (2.13) we know, therefore, that

$$\begin{aligned} l^2\psi &= a_l\psi \\ l_z\psi &= a_m\psi. \end{aligned} \tag{2.14}$$

A complete review of the mathematical exposition of angular momentum theory would now involve two discrete sections. One establishes the form of the *eigenvalues*, a_l and a_m, of (2.14) and the other describes the exact form of the *eigenfunctions*, ψ. Both pieces of theory involve no more physical input than has been described already and so their relegation to mathematical 'black boxes', in the spirit of this book, robs us of no physical insight. The student is, of course, asked to take the author's word for it: should he doubt it or wishes to see the algebraic nuts and bolts, he is referred elsewhere.[3,4] Here we confine ourselves to describing and understanding the form and relevance of the output.

2.4 *Eigenvalues and orbital quantum numbers*

First we consider the eigenvalues, a_l and a_m of (2.14). In words first, it turns out that these take the following form: The eigenvalues of l^2 comprise an integer, multiplied by the same integer, plus one—all times $\hbar^2$. For l_z, eigenvalues involve another integer—times $\hbar$, this time—whose value is related to the integer in the eigenvalue of l^2; ranging in integral steps from the positive to the equal negative value of the former. Thus:

$$\hat{l}^2\psi = \{\hbar^2 l(l+1)\}\psi: \qquad l = 0,\ 1,\ 2\ \ldots \tag{2.15}$$

$$\hat{l}_z\psi = \{\hbar\ m_l\}\psi: \qquad m_l = l,\ l-1,\ \ldots\ -l. \tag{2.16}$$

Hence for each value of l, there exist $(2l+1)$ values of m_l, ranging as shown in (2.16): we refer to the *orbital degeneracy*, $(2l+1)$. The table illustrates these relationships. It also shows how much more convenient it is to *label* the eigenfunctions by the numbers l and m_l rather than write them out in full each time. These eigenvalue labels are examples of *quantum numbers*. Let us be clear here:

$\hat{l}^2$ eigenvalue $l(l+1)\hbar^2$	Label l	Label m_l	Degeneracy $(2l+1)$
0	0	0	1
$2\hbar^2$	1	1, 0, −1	3
$6\hbar^2$	2	2, 1, 0, −1, −2	5
⋮	⋮	⋮	⋮

quantum numbers label eigenvalues but are *not* the eigenvalues themselves.

Problem Suppose we consider the state for some system in which $m_l = l$. Apparently the z component of angular momentum equals the total angular momentum. Therefore the x and y components are both zero. Hence we simultaneously know all three components of the angular momentum which is impossible because $\hat{l}_x$, $\hat{l}_y$ and $\hat{l}_z$ do not commute. Clarify this issue.

Given the background experience assumed here, it would be coy not to recognize the reader's familiarity with orbital quantum numbers and both the choice and form of symbols for the quantities in (2.15) and (2.16). However, it cannot be emphasized too often in this chapter that our discussion has been restricted from the beginning to one about the motion of an arbitrary particle—a half-brick perhaps—and is not tied in any way to the electrons in the atom. Of course, it is atoms we aim to study, but the point of the present discussion is to separate clearly the properties of angular momentum from others, like energy.

2.5 *The eigenfunctions—spherical harmonics*

The 'black box' mathematics that establishes the eigenvalue patterns for l^2 and l_z utilizes not only the theorem of commuting operators, as we have seen, but also the specific manner of non-commutation amongst all three components of **l** given in (2.12) together with a theorem about the expansion of functions which we discuss in other contexts later, in Chapter 6. Algebraic manipulations of a similar kind also furnish us with a complete description of the eigenfunctions of (2.15). These functions are called *spherical* *harmonics* and, being solutions of eigenvalue equations which relate solely to angular momentum, depend only upon angular displacements about the centre of rotation: they do not depend upon the distance of the particle from that centre. Therefore, a more natural system of coordinates than the rectangular or cartesian ones used so far is that of polar coordinates in which the separation of radial and angular functionality is made explicit.

In Figure 2.4 we recall the spherical polar coordinate scheme and its relationship to the cartesian one. Note, once again, the adoption of a right-handed cartesian frame and the definition of the polar angle ϕ to be positive when measured from the positive

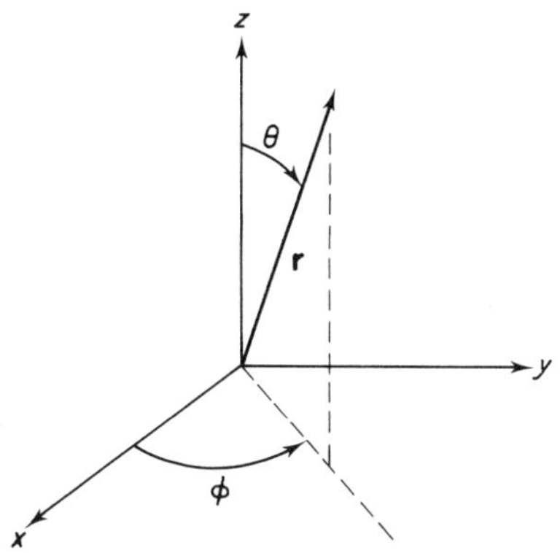

Figure 2.4 Standard definition of spherical polar coordinates with respect to a right-handed cartesian frame.

x-axis, towards or through the positive y-axis; that is, anticlockwise from x as viewed from the positive z axis towards the origin. These conventions accord with those discussed in connection with Figure 2.2 above. They are standard conventions and it is inadvisable to depart from them, even when the rotational sense in a given problem may seem irrelevant. In terms of these polar coordinates the angular momentum operators l^2 and l_z take the forms,

$$l^2 = -\hbar^2 \left\{ \frac{1}{\sin\theta} \frac{\partial}{\partial\theta} \left(\sin\theta \frac{\partial}{\partial\theta} \right) + \frac{1}{\sin^2\theta} \frac{\partial^2}{\partial\phi^2} \right\} \qquad (2.17)$$

$$l_z = -i\hbar \frac{\partial}{\partial\phi} \, . \qquad (2.18)$$

These should seem sensible really, at least so far as their general shape is concerned. The operator l_z refers to the angular momentum of an object with respect to the z-axis and so is connected solely with a rate of change of the rotational angle in the plane perpendicular to that axis—hence $\partial/\partial\phi$. Not surprising either is an observation that the operator for the total angular momentum involves both angular coordinates θ and ϕ.

Returning to the spherical harmonics, then, we write these functions of the two polar angular coordinates as $Y^l_m\,(\theta,\phi)$. Each function is labelled with the quantum numbers l and m_l but the latter is always written in the present context as m so as to avoid the typographical complexities of subscripted subscripts: confusion about the meaning never arises.†

The functions are so labelled because each pairing of quantum numbers corresponds to a unique spherical harmonic that simultaneously satisfies the eigenvalue equations of $\hat{l}^2$ and $\hat{l}_z$. Hence,

$$\hat{l}^2\, Y^l_m\,(\theta,\,\phi) = \{\hbar^2\, l(l+1)\}\, Y^l_m\,(\theta,\,\phi) \tag{2.19}$$

$$\hat{l}_z\, Y^l_m\,(\theta,\,\phi) = \{\hbar\, m_l\}\, Y^l_m\,(\theta,\,\phi)\,. \tag{2.20}$$

Specifically, $Y^2_1\,(\theta,\,\phi)$, for example, denotes an eigenfunction of a particle with total angular momentum squared of $6\hbar^2$ units and $\hbar$ units of angular momentum with respect to z as the quantization axis. Note here that both m_l and the spherical harmonics refer to the same quantization axis. This is so only by construction, definition and convenience: other schemes could have been established but would not correspond with standard practice.

Now a most interesting feature of angular momentum eigenfunctions is that the spherical harmonics each factorize into one part dependent only on the θ-coordinate, and one only on ϕ. We write

$$Y^l_m\,(\theta,\,\phi) = \Theta^l_{|m|}\,(\theta)\,\Phi_m\,(\phi) \tag{2.21}$$

and note at once that the occurrence of such a resolution of the two angular coordinates is by no means an obvious consequence of

† Different authors place these quantum labels in a variety of places—Y^l_m, Y^m_l, $Y_{l,m}$—and sometimes brackets are placed around the l in the first case $Y^{(l)}_m$. This author prefers Y^l_m or $Y^{(l)}_m$ so as to correspond with conventions in the theory of irreducible tensors—but it does not matter so long as it is clear.

what we have seen so far. It is the case, nevertheless, and we discuss the two parts in turn, beginning with the Φ function because it is the simpler.

Given the factorization of the spherical harmonics, we can readily determine the form of $\Phi(\phi)$ for ourselves by using the explicit form (2.18) of l_z in (2.20):

$$-i\hbar \frac{\partial}{\partial\phi} \left\{\Theta(\theta)\ \Phi(\phi)\right\} = m_l\hbar \left\{\Theta(\theta)\ \Phi(\phi)\right\} \tag{2.22}$$

or, since $\partial/\partial\phi$ does not operate on any function which is independent of ϕ,

$$\frac{\partial}{\partial\phi} \Phi(\phi) = im_l\ \Phi(\phi). \tag{2.23}$$

The general solution of this equation is

$$\Phi_m(\phi) = \Phi_m(\phi+2\pi)\ ; \tag{2.24}$$

that is,

$$e^{im\phi} = e^{im(\phi+2\pi)} = e^{im\phi}\ e^{2\pi im} \tag{2.25}$$

and hence that $e^{2\pi im} = 1$ which is true only for m integer—but we know that already. We get the normalization factor by integration as usual:

$$\int_0^{2\pi} \Phi_m^*\ (\phi)\ \Phi_m(\phi)\ d\phi = N^2 \int_0^{2\pi} d\phi = 1\ , \tag{2.26}$$

whence N is $(2\pi)^{-1/2}$. Altogether, therefore, the Φ function in (2.21) is just

$$\Phi_m(\phi) = \frac{1}{\sqrt{2\pi}}\ e^{im\phi} \tag{2.27}$$

and we observe that these functions are always complex except if $m = 0$, when $\Phi \rightarrow 1/\sqrt{2\pi}$.

The $\Theta^l_{|m|}(\theta)$ functions have a name. They are called the *associated Legendre polynomials*. When normalized to unity they take the form:

$$\Theta^l_{|m|}(\theta) = \frac{(-1)^l}{2^l l!}\sqrt{\frac{2l+1}{2}\frac{(l-|m|)!}{(l+|m|)!}}\,\sin^{|m|}\theta\,\frac{d^{l+|m|}\,(\sin^{2l}\theta)}{(d\cos\theta)^{l+|m|}}\,. \qquad (2.28)$$

I present this fearsome expression only to show that the 'black box' algebra already alluded to, actually defines the eigensolutions of the angular momentum operators completely. This is an analytical expression which, as we shall see in a moment, boils down to remarkably simple trigonometric functions of θ. The factorials in the general expression emerge from repeated integrations required by recursion-type solutions of certain differential equations. Observe that the quantum number m ($\equiv m_l$) only occurs as its modulus which is why the Θ function is so labelled. These normalized associated Legendre polynomials are intrinsically positive, despite the occurrence of the factor $(-1)^l$ in (2.28).

Let us see what these functions look like for a couple of specific cases. The simplest is $\Theta^0_0(\theta)$:

$$\Theta^0_0(\theta) = \frac{1}{1}\sqrt{\frac{1}{2}}\,\sin^0\theta\,\frac{d^0\sin^0\theta}{(d\cos\theta)^0} = \sqrt{\frac{1}{2}}\,. \qquad (2.29)$$

Only slightly more complicated is the evaluation of $\Theta^1_0(\theta)$:

$$\Theta^1_0(\theta) = -\frac{1}{2}\sqrt{\frac{3}{2}\cdot\frac{1}{1}}\,\sin^0\theta\,\frac{d\sin^2\theta}{(d\cos\theta)} \qquad (2.30)$$

and by making the substitution $x = \cos\theta$ in the last factor we find that

$$\frac{d\sin^2\theta}{(d\cos\theta)} = -2\cos\theta, \qquad (2.31)$$

whence

$$\Theta^1_0(\theta) = \sqrt{\frac{3}{2}}\cos\theta\,. \qquad (2.32)$$

Table 2.1 Some associated legendre polynomials.

l	m_l	$\Theta^l_{\lvert m \rvert}(\theta)$
0	0	$1/\sqrt{2}$
1	0	$\sqrt{3/2}\cos\theta$
1	± 1	$\sqrt{3/4}\sin\theta$
2	0	$\sqrt{5/8}\,(3\cos^2\theta - 1)$
2	± 1	$\sqrt{15/4}\sin\theta\cos\theta$
2	± 2	$\sqrt{15/16}\sin^2\theta$

Table 2.1 lists a few more normalized associated Legendre polynomials that will be useful to us later on.

2.6 *The shapes of some spherical harmonics*

Simplifying the algebra of angular wavefunctions is all very well but it is rather more revealing to draw one or two of them on paper. Generally, of course, the spherical harmonics are complex because of the complex exponential in the Φ function of (2.27); but, as noted above, these factors collapse to a real constant in those cases when $m_l=0$. Therefore all $Y^l_0\,(\theta,\phi)$ functions are figures of revolution with respect to the z-axis, and real. We shall look at the first few of these.

The simplest of all, Y^0_0 is simply a constant, being independent of θ (from 2.29) and of ϕ (from 2.27). The shape of an angular function whose magnitude does not change as we cover all angular space is just a sphere and, as the sign of the function is positive, we draw it as in Figure 2.5(a).

To get $Y^1_0\,(\theta,\,\phi)$ we first examine the behaviour of the $\cos\theta$ of (2.32) in the polar coordinate scheme. We plot this as shown in Figure 2.6(a). As θ takes values 0, 45°, 90°, the magnitude of the function is $1, 1/\sqrt{2}, 0$: for $\theta = 135°$, 180°, $\cos\theta$ takes magnitudes $1/\sqrt{2}$, 1. While the magnitudes of the functions in the second quadrant repeat those in the first, the signs change from positive to negative. Further it is not difficult to show that the locus traced out in this way by $\cos\theta$ comprises a pair of exact semicircles. Ignoring the normalizing constants, the spherical harmonic $Y^1_0\,(\theta,\,\phi)$ that we require is then obtained by rotating the polar plot of $\cos\theta$ about the z-axis to obtain the cylindrical symmetry of the $\Phi(m_l=0)$ function—

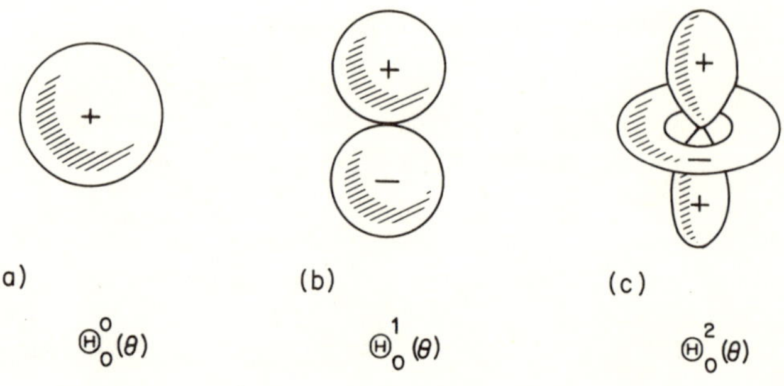

Figure 2.5 Three associated Legendre polynomials.

and we get the shape shown in Figure 2.5(b), comprising two spheres of opposite parity.

Turning now to $Y_0^2(\theta, \phi)$ we refer to Figure 2.5(c) illustrating the construction of $(3\cos^2\theta - 1)$ in the polar coordinate scheme. The locus is no longer a combination of simple circular parts, and the parity changes twice. The figure is symmetrical in both quadrants with respect to both magnitude and sign and we note the nodal plane at the angle $\cos^{-1}\sqrt{1/3}$ in the upper quadrant. When compounded with the constant Φ function, there results the second-order spherical harmonic we seek, shown in Figure 2.5(c).

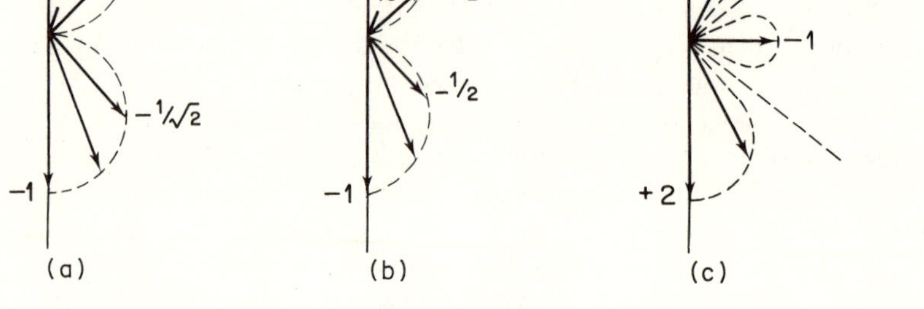

Figure 2.6 Polar plots of (a) $\cos\theta$, (b) $\cos^2\theta$, and (c) $3\cos^2\theta - 1$.

2.7 *The relationship with atomic orbitals*

The shapes of the functions in Figure 2.5 are well known to us, of course. They *appear* to be the *s*, p_z, d_z^2 orbitals with which we are so familiar—but we must be careful! Orbitals are to do with electrons and atoms, whereas the angular momentum theory we have studied here referred (repeatedly) to the motion of an *arbitrary* object: for that reason at least, we know that we cannot have really constructed orbitals. The point here, of course, is that while the spherical harmonics are functions only of the angular coordinates θ and ϕ, orbitals involve the radial coordinate *r* also. Indeed, orbitals are eigenfunctions of a one-electron Hamiltonian operator—an energy operator—and their associated eigenvalues are energies, whereas the eigenvalues of spherical harmonics are angular momenta.

It is common to introduce quantum chemistry via the Schrödinger equation for the hydrogen atom and the reader will be familiar with the form of the resulting hydrogen wavefunctions as simple products of radial and angular parts. The radial functions arise from a consideration of energy via the Hamiltonian operator, as discussed in Chapter 1. The angular parts, though often introduced without explicit reference to angular momentum, are just the functions we have been describing in the last section. They have to be, of course, because that theory referred to the angular motion of *any* object and an electron in an atom is just one particular example of that. In other words, the orbitals of hydrogen—or of any other atom—*must* have the angular shapes we refer to as *s, p, d* . . . simply because the electrons they describe are objects possessing angular momentum. The shapes of *s, p, d* . . . functions are intrinsic manifestations of angular momentum and isotropic free space rather than properties of bound electrons in atoms *per se*. Indeed, when we write an atomic orbital as

$$\psi(r, \theta, \phi) = R_{nl}(r)\, Y^l_m(\theta, \phi), \qquad (2.33)$$

remember that the specific forms of the radial function *R*(*r*) depends upon the Hamiltonian operator and so varies as we direct our attention from hydrogen to atoms of other elements.

A further point arises here, to do with the manner of drawing orbitals. Generally throughout mainstream chemistry we sketch shapes for orbitals like those in Figure 2.5 in a rough-and-ready way merely to indicate their general directional nature and overlap possibilities. For *p* orbitals especially, the exact shapes that appear in books or on blackboards vary, depending, apparently, upon the taste of, and artistic licence taken by, the drawer. Usually we can be relaxed about these variations but we should be aware, nevertheless, that strictly there are right and wrong ways. Taking the p_z function to illustrate these, we note once more that the exact shape of $Y_0^1(\theta, \phi)$ comprises the two spheres in Figure 2.5(b). However, the square of this function, which we construct as in Figure 2.6(b) is the more directed entity shown in Figure 2.7(b). The radial parts $R(r)$ of some hydrogen atomic functions are shown in Figure 2.8 and the corresponding radial probability functions $r^2R^2(r)$ in Figure 2.9. Since the total electron density distribution for a particular atomic orbital involves the square of the orbital, we get that for the $2p_z$ hydrogen orbital, say, by multiplying the functions sketched in Figure 2.7(b) by the squared radial quantity shown for 2*p* in Figure 2.9. The resultant density function is shown in Figure 2.7(c): note that it coincidentally approximates the double sphere shape of Figure 2.7(a) again, but this is just because it is those parts of the shape in Figure 2.7(b) at middling distances that are multiplied by the radial factor where it peaks. Compare now Figure 2.7(d), showing the total wavefunction (2.33) for a hydrogen $3p_z$ orbital (unsquared) and in Figure 2.7(e) for the corresponding electron density function. It is clear that conventional 'back-of-envelope' sketches often omit any reference to the possible nodal features of the radial functions.

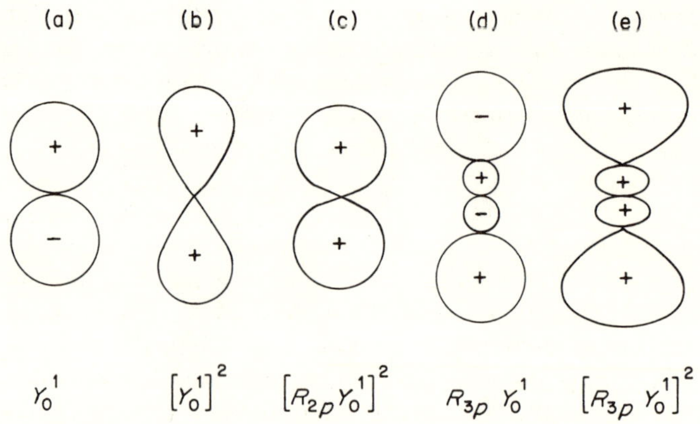

Figure 2.7 Various schematic drawings of '*p* orbitals'.

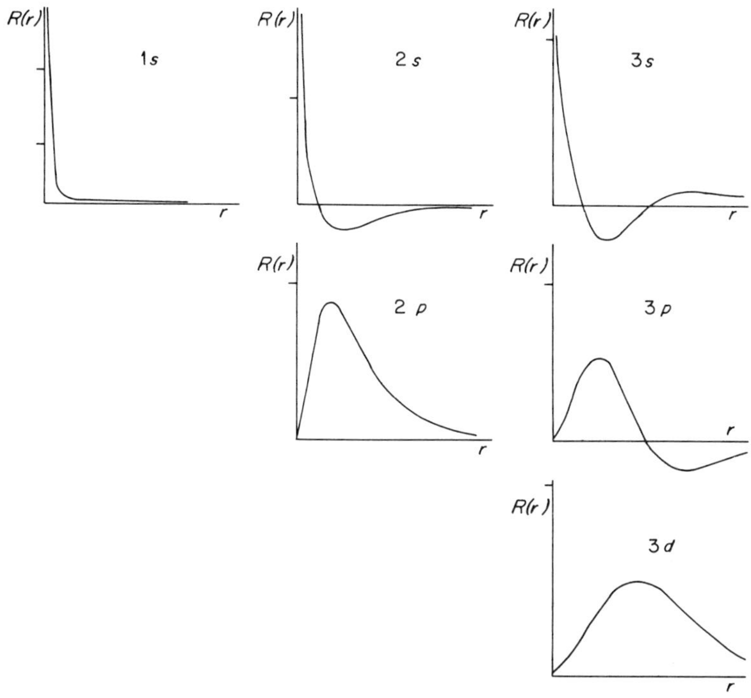

Figure 2.8 Radial parts of hydrogenic wavefunctions (arbitrary scales).

2.8 *Electron spin*

In noting the 'rules of the quantum-mechanical game' in Chapter 1 and by applying the operator correspondences to the classical equations of angular momentum in this chapter, the reader is hopefully a little more practised in what is, after all, a very big intellectual leap in the process of understanding twentieth-century science. He or she will already be acquainted with the concepts of electron spin on an *ad hoc* basis. We may *assert* that electrons, for example, have three *intrinsic* properties—rest-mass, charge, and spin; or, if you prefer, rest mass, charge and a magnetic moment. It does little good to ask why. One might similarly enquire into the connection between the properties of an orange as being round, of orange colour and having a characteristic orange taste. (Electronic *orbital* angular momentum, by the way, is not an intrinsic property, but an *extrinsic* one, being manifest by the electron in its environ-

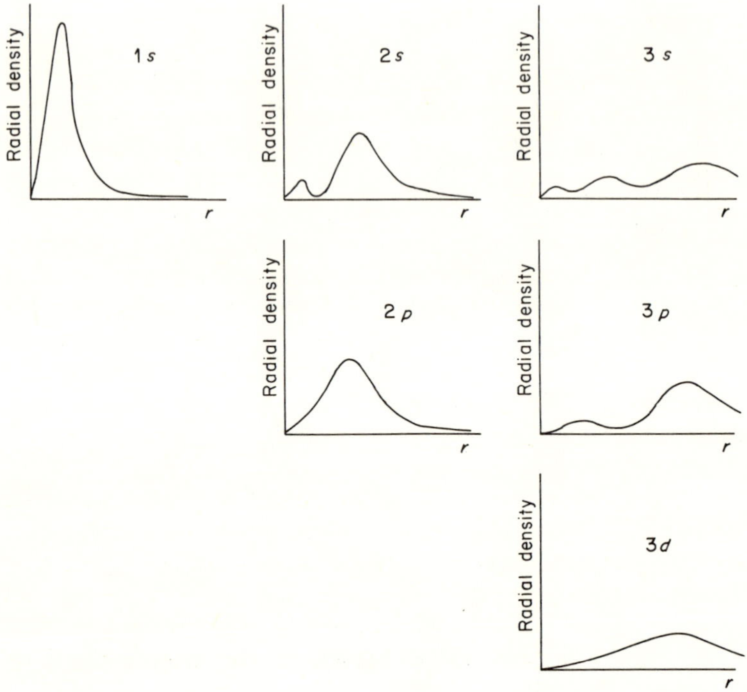

Figure 2.9 Radial probability distributions associated with hydrogenic wave functions (arbitrary scales).

ment.) Returning to the question of spin; given that 'explanation' means 'relationship with other, perhaps more familiar (current or utilized) ideas', it can be said that electron spin arises from the second major building block of this century's science—the theory of relativity. It is my belief that many students will, at this stage, have some knowledge of the special theory of relativity however overcast with disquiet at the mathematical content. For them, I present a 'chatty' discussion of the connection between spin and relativity. Others should note that what follows is not essential for an ability to manipulate spin quantum numbers in chemistry and that useful understanding can begin at any convenient point.

Within the special theory of relativity one is much concerned with transformations in 'space-time'. Here time and ordinary three-dimensional space appear on almost equivalent footings describing a four-dimensional continuum within which 'events' (as opposed to locations or time instants) are plotted. For us it is important merely to recognize that the space–time continuum is not the construct of another world but is wholly of this one: our everyday unawareness of it derives from the fact that for most physical observations, space and time variables are nearly perfectly uncoupled such that equations containing both sets of variables can be factorized almost exactly in that way. Now an essential feature of any *description* of the world, or a part of it, is that the predicted values of those quantities which are experimentally observable must not depend upon the choice of coordinate frame. A trivial but graphic example is to note that an equation describing the heat radiated by an electric fire had better give the same answer whether the cold observer is standing on his head or his feet. This is a centrally important notion, much utilized in theoretical chemistry, for example, and will arise again in the present book. As a proper description of the real world—so far as is currently understood—requires a simultaneous consideration of space and time on equivalent footings, all descriptions of physical observables should be invariant to appropriate four-coordinate transformations called Lorenz transformations. The same requirement holds for descriptions generated by quantum mechanics too. But consider the form of the time-dependent Schrödinger equation:

$$\left\{-\frac{\hbar^2}{2m_e}\left(\frac{\partial^2}{\partial x^2}+\frac{\partial^2}{\partial y^2}+\frac{\partial^2}{\partial z^2}\right)-i\hbar\frac{\partial}{\partial t}+V\right\}\psi(x, y, z, t)=0 \tag{2.34}$$

The space and time variables do not appear in the same way—the powers of $\{x, y, z\}$ and t are different. It follows immediately that this equation cannot transform invariantly under any Lorenz transformation of the space–time continuum whose transformation rules involve all four coordinates in the same way. While this was recognized in the early days of quantum mechanics, all attempts to reformulate quantum ideas in a relativistically invariant way failed until Dirac forged a way through, and in so doing discovered several known and unknown consequences of immense importance to the subject. In essence, he produced a Schrödinger-like equation which simultaneously satisfied the postulates of special relativity and of quantum theory but certain features of his equation describes properties of the electron which had no parallel within Schrödinger's theory. First, the eigenfunctions of Dirac's equation represented two sorts of condition for the electron that came to be referred to as particles with the names electron and positron. Secondly, with each of these particles was associated an angular momentum property additional to that already dealt with by Schrödinger's equation (and our earlier discussion). This extra property turned out to possess precisely those properties of spin angular momentum that had been postulated *ad hoc* from earlier painstaking study of atomic spectra. In passing, remember that the story being told in the first few chapters of this book was first built by 'reverse logic', so to speak, from the splendid observational edifices created by the atomic spectroscopists—a magnificent intellectual achievement. The bottom line for us in this brief sketch of Dirac's contribution must be to reiterate that electronic spin angular momentum is a consequence of the simultaneous application of quantum mechanics and special relativity. Although we shall have occasion to refer back to this 'thumb-nail' summary, it is appropriate now that we step back to the level of formal description of the spin properties for which we need not know about its origins.

Without proof, therefore, we introduce the spin analogues of the orbital angular momentum operators l^2 and l_z. Eigenfunctions of the spin angular momentum operators s^2 and s_z satisfy equations analogous to (2.15) and (2.16), namely,

$$\hat{s}^2 \psi_s = \{\hbar^2 s(s+1)\} \psi_s \tag{2.35}$$

$$\hat{s}_z \psi_s = \{\hbar m_s\} \psi_s. \tag{2.36}$$

The structural similarities shown by the spin and orbital eigenvalue equations ensure, as we shall see, near one-to-one parallels between the ways we manipulate spin and orbital quantum numbers. There are differences between the two sets of equations, however. As for the orbital case, we refer to s as a spin quantum number labelling the eigenvalue of $\hat{s}^2$ appearing in the braces. Yet, while l takes any integer value, s for a single electron takes only the one value, ½. Note that the s value and all l values are intrinsically positive, for they label eigenvalues of total spin or orbital angular momentum *squared*. The spin magnetic quantum number m_s labels eigenvalues of the operator referring to the component of spin angular momentum parallel to the z-axis, and values for it follow the pattern of those for m_l, in that m_s ranges $+s$ to $-s$ in integer steps. For the single electron, therefore, m_s makes just two values,

$+\frac{1}{2}$ and $-\frac{1}{2}$. These are sometimes labelled as α and β respectively, or by upward and downward pointing arrows—$\uparrow$, $\downarrow$. The spin degeneracy is given by the general formula $(2s+1)$, analogous to the orbital degeneracy $(2l+1)$. In the case of a single electron, we have $(2s+1) = (2.\frac{1}{2}+1) = 2$, so the formula works properly with half-integral values.

Earlier, we wrote the eigenfunctions of l^2, l_z as ψ, and then $Y^l_m(\theta, \phi)$, while ψ_s represents a function of spin coordinates in equations (2.35) and (2.36). Though not essential, it is convenient to use Dirac's bra-ket notation here. We characterize the eigenfunctions we require by reference simply to the quantum numbers for orbital and spin angular momentum; or to a subset of these if the context allows a briefer description without confusion. For example, (2.15), (2.16), (2.35), (2.36) can be written in any of the following ways,

$$\hat{l}^2 \mid l\ m_l\ s\ m_s \rangle = \hbar^2\ l(l+1)|l\ m_l s\ m_s\rangle \tag{2.37}$$

$$\hat{l}_z \mid m_l \rangle = \hbar\ m_l|m_l \rangle \tag{2.38}$$

$$\hat{s}^2|l m_l s m_s\rangle = \hbar^2\ s(s+1)|l m_l s m_s\rangle \tag{2.39}$$

$$\hat{s}_z|s m_s\rangle = \hbar m_s|s\ m_s\rangle\ . \tag{2.40}$$

In the special case of the spin angular momentum of single electrons we can rewrite the last two equations as

$$\hat{s}^2|\alpha\rangle = \tfrac{3}{4}\ \hbar^2|\alpha\rangle \qquad \hat{s}^2|\beta\rangle = \tfrac{3}{4}\ \hbar^2|\beta\rangle \tag{2.41}$$

$$s_z|\alpha\rangle = \tfrac{1}{2}\ \hbar|\alpha\rangle \qquad \hat{s}_z|\beta\rangle = -\ \tfrac{1}{2}\ \hbar|\beta\rangle. \tag{2.42}$$

In view of the singular value of s for an electron, these last two relationships summarize all about the electron spin we need to know so it may seem unnecessarily complicated to write the more general forms of (2.35), (2.36) or (2.39), (2.40). Reasons for doing so will become clear later.

Finally, note that a complete description of the spin and orbital angular momentum of an electron or atom requires specification of four quantum numbers—l, m_l, s, m_s—and not just three (orbital, spin, magnetic) as is occasionally inferred from some general chemistry texts.

CHAPTER 3

Many-Electron Atoms

3.1 *Angular momentum operators and commutators*

As advertised in the latter part of Chapter 1 we are aiming towards a description of eigensolutions that characterize not only the behaviour of atoms possessing a single electron outside of closed shells but also of those with groups of interacting valence electrons. While a complete solution of the Schrödinger equation for one-electron atoms is possible but difficult, similar solutions for so-called many-electron atoms appear to be impossible analytically and quite hard even by numerical methods on a computer. Such problems actually have been solved for most atoms but as with all such calculations, presentation of numerical eigenvalues does little to further understanding of principles. What we are able to do here for ourselves is to characterize the eigensolutions of many-electron atoms with respect to angular momentum and in the process learn something of the nature of electron–electron interactions in atoms. It is useful, though not essential, to hold an example of a many-electron atom at the back of our minds and for this we shall consider a transition metal with a $3d^2$ configuration outside of closed shells. The choice of the d^2 system is made on the grounds that it has relevance to a whole body of discussions which take place in a subject called ligand-field theory that the student will deal with later and elsewhere;[5–7] that it provides enough complexity to be interesting and illustrate the theoretical features we shall study; and yet it is not so complex as to obscure the important points with technicalities.

The various electrons of interest to us in a many-electron atom execute coupled motions of some kind. That is to say that they interact in various ways, the most obvious kind of interaction or coupling being the electrostatic repulsion expected between two similarly charged particles. At any given instant, each electron possesses some amount of angular momentum—total or z component, orbital or spin. However, while these quantities are constant for uncoupled electrons, they vary with time when electrons interact. Let us illustrate the idea with an analogy to do with oscillating pendula.

Consider the pair of isolated pendula in Figure 3.1(a). Each may be excited to oscillate with a characteristic frequency dependent upon its length which can be determined at any time simply by timing the separate motions over some fixed time interval. However, if the pendula interact, which we can arrange here by connecting them with a light spring as in Figure 3.1(b), then we would observe that neither pendulum oscillates with an unchanging frequency: they are not characterized by constants of their individual motion. Indeed, if we observe them long enough—for a time depending in part on the difference of their free frequencies—the situation will arise when one pendulum acquires a vanishing amplitude while the other's becomes maximal: at a later time still, the pendula will exchange roles in this regard. One constant of motion we *can* discern, however, is the frequency of exchange of energy between the two pendula. This is a property of the coupled *pair*, of course, rather than of the individuals. It is typical of all coupling processes that characteristics of the individuals are sac-

Figure 3.1 Two pendula. (a) Independent, uncoupled, free with separate identifiable constants of motion. (b) Coupled (by a light spring) with no separate constant frequencies.

rificed to the ensemble. The other factor that determines the rate at which our pendula exchange their energy is the stiffness of the spring connecting them. If the spring is very weak, the energy exchange process will be very slow and an observation of the oscillations of either pendulum over a modest time interval would yield a fairly constant motion. Generalizing this idea we infer that descriptions of the dynamics of the individuals in a weakly coupled set of systems can yield characteristic constants of motion to a fair, if variable, degree of accuracy. We shall refer back to this idea later on.

A major point of this little analogy is to make clear that the causes and consequences of the electronic coupling process in atoms need not be related directly otherwise. As a result, for example, of interelectron repulsion, some angular momenta of individual electrons are no longer constants of motion and we must now consider the characteristics of angular momentum of the electron ensemble as a whole. Other consequences of such electronic coupling concern ensemble *energies*, of course, but, apart from a discussion of some general qualitative rules on the matter, these quantities are hard to calculate, methods for which rely totally upon numerical techniques, and do not fall within our field of study. So we shall concentrate on angular momentum.

We begin by constructing angular momentum operators for groups of atomic electrons. For this we recognize that the total electronic angular momentum of the system at any instant is given simply by the vectorial sum of the momenta for each electron and so we define collective or ensemble operators as

$$L_x = l_x(1) + l_x(2) + \ldots \quad (3.1)$$

$$L_y = l_y(1) + l_y(2) + \ldots \quad (3.2)$$

$$L_z = l_z(1) + l_z(2) + \ldots \quad (3.3)$$

so adopting from now on the nomenclature that *capital letters refer to many-electron systems; lower-case ones to one-electron species*. Analogous to (2.9), we have, for the total orbital angular momentum squared, the operator

$$L^2 = L_x^2 + L_y^2 + L_z^2 \quad (3.4)$$

Spin angular momentum operators are defined similarly:

$$S_x = s_x(1) + s_x(2) + \ldots \tag{3.5}$$

$$S_y = s_y(1) + s_y(2) + \ldots \tag{3.6}$$

$$S_z = s_z(1) + s_z(2) + \ldots \tag{3.7}$$

$$S^2 = S_x^2 + S_y^2 + S_z^2 \tag{3.8}$$

Now we consider the commutation rules of these operators, amongst themselves and with respect to the one-electron components, and to the atomic Hamiltonian. Some commutators are trivially obvious. Since spin operators act exclusively upon the spin parts of the angular momentum wavefunctions and orbital operators only upon the space parts, all spin operators commute with all space operators. Similarly, since any one-electron angular momentum operator acts only upon the coordinates of the electron in question—that is, $l_x(1)$ acts on electron 1, $l_y(2)$ only upon electron 2—the various angular momenta of one electron commute with all those of any other electron. Equally obvious is that all spin operators commute with the Hamiltonian because that operator does not involve any spin coordinates.

It is only slightly more difficult to show that the (upper case) orbital angular momentum operators commute with the Hamiltonian also. We begin with L_z, which in polar coordinates is

$$L_z = -i\hbar\left(\frac{\partial}{\partial\phi_1} + \frac{\partial}{\partial\phi_2} + \ldots\right), \tag{3.9}$$

where $\{\phi_i\}$ are the polar ϕ angles of the various electrons. Now the Hamiltonian $\mathcal{H}$ of any atom involves the ϕ angles in two parts. First, in the kinetic energy term involving the Laplacian operator, $\nabla^2 \equiv \partial^2/\partial x^2 + \partial^2/\partial y^2 + \partial^2/\partial z^2$, for each electron—which in spherical polar coordinates is written

$$\nabla^2 = \frac{1}{r^2}\frac{\partial}{\partial r}\left(r^2\frac{\partial}{\partial r}\right) + \frac{1}{r^2\sin\theta}\frac{\partial}{\partial\theta}\left(\sin\theta\frac{\partial}{\partial\theta}\right) + \frac{1}{r^2\sin^2\theta}\frac{\partial^2}{\partial\phi^2} \tag{3.10}$$

—we note that the only part involving ϕ is of the form $\partial^2/\partial\phi^2$ for each electron: using similar arguments to those given above, it is clear that L_z of (3.9) commutes with ∇^2 of (3.10). Consider now the potential energy terms in $\mathcal{H}$ to do with the Coulomb operator $\Sigma_{i<j}\, e^2/r_{ij}$. This involves a sum of terms proportional to $1/r_{12}^2$. Reference to Figure 3.2 immediately shows that the instantaneous vector $\mathbf{r}_{12}$ between two electrons concerns the angle ω between the two vectors $\mathbf{r}_1$ and $\mathbf{r}_2$ via the cosine formula

$$r_{12} = (r_1^2 + r_2^2 - 2r_1r_2\cos\omega)^{1/2} \tag{3.11}$$

and hence involves ϕ_1 and ϕ_2 only in the combination $(\phi_1 - \phi_2)$. Operating with L_z on e^2/r_{12} thus yields, on differentiation by parts,

$$\begin{aligned} &-i\hbar\left(\frac{\partial}{\partial\phi_1} + \frac{\partial}{\partial\phi_2} + \ldots\right)\frac{e^2}{r_{12}} \\ &= -i\hbar e^2\,\frac{\partial(e^2/r_{12})}{\partial(\phi_1-\phi_2)}\left\{\frac{\partial(\phi_1-\phi_2)}{\partial\phi_1} + \frac{\partial(\phi_1-\phi_2)}{\partial\phi_2}\right\} + \ldots \\ &= 0 \end{aligned} \tag{3.12}$$

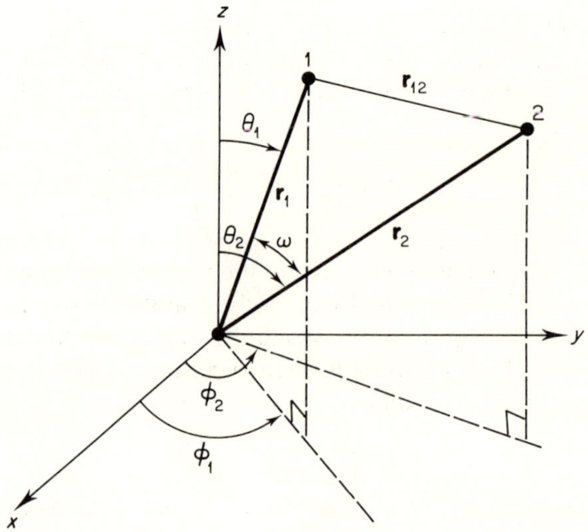

Figure 3.2 The angle ω is a scalar and independent of the reference frame.

Hence L_z commutes with H and, since atomic Hamiltonians have spherical symmetry, L_x and L_y must similarly commute with H; as also L^2.

Finally, making use of the one-electron commutation relationship (2.12), together with the various commutators established above, we can show that

$$[L_x, L_y] = i\hbar\, L_z \tag{3.13}$$

$$[L_y, L_z] = i\hbar\, L_x \tag{3.14}$$

$$[L_z, L_x] = i\hbar L_y, \tag{3.15}$$

as for the one-electron relationships themselves.

3.2 *Terms and term symbols*

One can summarize our exploration of the commutators for many-electron angular momenta operators by observing a very close parallel with the starting point constructed in the single-particle case in Chapter 2. As before, we select a pair of commuting orbital angular momentum operators—conventionally and obviously, L^2 and L_z—and make use of the theorem in (2.13) and all that followed. Hence, in words first: for L^2 the eigenvalues are characterized by an integer multiplied by that same integer plus one, times $\hbar^2$; while for L_z, eigenvalues are given by another integer times $\hbar$. The relationships between these two integers are precisely the same as before, because the algebra derives from exactly parallel commutation rules. They are determined, in short, by the properties of angular momentum and, in this context, a many-electron atom exemplifies these concepts just as well as a one-electron atom—or a half-brick! So we write the same equations as (2.15) and (2.16) but shift case:

$$\hat{L}^2\,|L, M_L\rangle = \{\hbar^2 L(L+1)\}\,|L, M_L\rangle \tag{3.16}$$

$$\hat{L}_z\,|L, M_L\rangle = \{\hbar\, M_L\}\,|L, M_L\rangle, \tag{3.17}$$

where $L = 0, 1, 2 \ldots$ and $M_L = L; L-1 \ldots -L$ as before. The orbital degeneracy of an eigenfunction characterized by the quantum number L is $(2L + 1)$.

We can write virtually the same equation for the spin-angular momenta of a many-electron atom:

$$\hat{S}^2 \,|S, M_S\rangle = \{\hbar^2 \, S(S+1)\} \,|S, M_S\rangle \tag{3.18}$$

$$\hat{S}_z \,|S, M_S\rangle = \{\hbar \, M_S\} \,|S, M_S\rangle \,. \tag{3.19}$$

This time, however, the values taken by the total spin quantum number S form integer-separated series based *either* on zero or on one-half:

$$S \begin{cases} \rightarrow 0, 1, 2 \ldots \\ \text{or} \\ \rightarrow \tfrac{1}{2}, \tfrac{3}{2}, \tfrac{5}{2} \ldots \end{cases} \tag{3.20}$$

We shall see why this happens in the next section: for the moment, recall that the lower case $s = \tfrac{1}{2}$ for a single electron failed to parallel l also. However, the spin-degeneracy $(2S+1)$ follows the pattern, for M_S values range $S, S-1 \ldots . -S$.

We summarize the position reached so far by Table 3.1, about which some further remarks may be made concerning definitions and jargon. The eigenfunctions of one-electron Hamiltonian operators (for atoms *or* molecules, actually) describe *orbitals*. To repeat the point but in reverse: *orbitals are one-electron wavefunctions*. The complete description of any orbital requires a specification of both spin and space character. As space and spin coordinates are independent—at least so far as our discussion has so far revealed—an orbital may be written as a simple product of space-only and spin-only functions, ϕ_o and ϕ_s respectively:

$$\phi \equiv \phi_o \, \phi_s. \tag{3.21}$$

An orbital whose spin and space parts are both described is called a *spin-orbital*. For example, what in mainstream chemical parlance might be written as $\uparrow\!\downarrow$, comprises a pair of spin-orbitals $\phi\alpha$ and $\phi\beta$. In answer to the question, 'How many electrons may be put into an orbital?' the correct response is 'one'; because (again) orbitals are one-electron wavefunctions. Of course, we need not be pedantic in the context of everyday chemical discussion, but here it is important to get it right. Strictly, then, there is no question of 'putting electrons into' an orbital, for this latter is not defined in the absence

Table 3.1

One-electron atoms $h\phi = \epsilon\phi$	Many-electron atoms $H\psi = E\psi$
$l^2\phi = \hbar^2 l(l+1)\phi$ $l_z\phi = \hbar\, m_l\phi$ $l \rightarrow$ 0, 1, 2, 3, 4, 5 ... s p d f g h ... $m_l \rightarrow$ 0 1 2 0 1 ... −1 0 −1 −2 Degeneracy $(2l+1)$ 1, 3, 5, 7, 9, 11 ... $m_l \rightarrow l, l-1 \ldots -l$	$L^2\psi = \hbar^2 L(L+1)\,\psi$ $L_z\psi = \hbar\, M_L\,\psi$ $L \rightarrow$ 0, 1, 2, 3, 4, 5 ... S P D F G H $M_L \rightarrow$ 0 1 2 0 1 ... −1 0 −1 −2 Degeneracy $(2L+1)$ 1, 3, 5, 7, 9, 11, ... $M_L \rightarrow L, L-1 \ldots -L$
$s^2\phi = \hbar^2\, s(s+1)\phi$ $s_z\phi = \hbar\, m_s\phi$ $s = \frac{1}{2}$ only $m_s = +\frac{1}{2}, -\frac{1}{2}$ or α, β or $\uparrow, \downarrow$ Degeneracy $(2s+1) = 2$ $m_s = s, s-1 \ldots -s$	$S^2\psi = \hbar^2\, S(S+1)\psi$ $S_z\psi = \hbar\, M_S\,\psi$ $S \rightarrow 0, 1, 2 \ldots$ or $\frac{1}{2}, \frac{3}{2}, \frac{5}{2} \ldots$ Degeneracy $(2S+1)$ 1, 3, 5 ... or 2, 4, 6 ... $M_S = S, S-1 \ldots -S$
Orbitals or one-electron wavefunctions	Term wavefunctions or many-electron wavefunctions

of the electron in the system for which the orbital is an eigenfunction.

The eigenfunctions of many-electron atoms are called *term wavefunctions* (at the present stage of presentation—but see Chapter 4). In reply to 'How many electrons can be put into a term wavefunction?' the correct answer is 'Electrons are not "put into" a wavefunction: each term wavefunction of an *n*-electron problem describes the behaviour of exactly *n* electrons'. There is no question of drawing a horizontal line with a series of upward or downward pointing arrows on it. Each two-electron wavefunction arising after the Coulomb interactions in a d^2 configuration, for example, relates to the various behaviours—the various angular momenta—of exactly two electrons.

The angular momenta of a many-electron atomic wavefunction can be represented sufficiently by the spin and orbital quantum numbers: we write† the ket $|L\,S\,M_L\,M_S\rangle$ to describe eigenvalues with respect to the four operators, L^2, S^2, L_z, S_z, respectively. Now orbital- and spin-degeneracies are necessarily complete; that is, are $(2L + 1)$-fold orbitally degenerate and $(2S + 1)$-fold spin degenerate. That these degeneracies *are* complete follows from the isotropy of space. Thus, if the total orbital angular momentum squared of an atom in a given situation is given by $\hbar^2L(L+1)$ units, a complete range of experimental viewpoints is always possible and between them will record values for the z components of angular momentum as $\hbar L$, $\hbar(L-1)$ . . . $-\hbar L$. Consider a classical analogue of an object observed to be rotating clockwise as seen by one observer but anti-clockwise by another who faces the first. The total angular momentum squared is necessarily a positive quantity and seen to be the same by both observers, despite their different conclusions regarding the angular momentum relative to some space-fixed axis. Altogether, therefore, the orbital degeneracy of a function specified by the quantum number L is $(2L+1)$ and by S is $(2S+1)$. Space and spin properties are independent of one another and so the total degeneracy is just given by the product, $(2L+1)(2S+1)$. All this can be summarized by writing $|L, S\rangle$ to represent what is called a *term*. A term comprises $(2L + 1) \times (2S + 1)$ term wavefunctions whose M_L and M_S values span the L and S values by ranging from the positive to the negative of these values in integral steps. So a term is a set of many-electron wavefunctions. For historical reasons, which do not help our understanding particularly, a term is not represented by $|L, S\rangle$ but by a *term symbol*, written as

$$^{2S+1}L\,. \tag{3.22}$$

Except for the historical development, these symbols do not appear particularly logical, for the orbital quantum number is stated while spin is described through the spin degeneracy. However, the use of these symbols is thoroughly established, so we must follow the convention. The value of L is represented by a letter, as given in Table 3.1. The first four—*S, P, D, F*—stood for sharp, principal, diffuse and fundamental in the original literature of atomic spectroscopy. For us the names are merely confusing and, for the purposes of quantum chemistry at least, are best forgotten. After these first

† The order of the quantum numbers is not important—some authors prefer $|L\,M_L\,S\,M_s\rangle$ for example—check the conventions in other texts.

four letters, the sequence follows the alphabet, except that *J* is omitted, for reasons that will become clear later. Examples of term symbols then, are ^3F, ^2D, ^5H, read as 'triplet *F*', 'doublet *D*', 'quintet *H*', and signify sets of many-electron wavefunctions (the number is *not* indicated by the term symbol) with *L*, *S* values 1, 3; ½, 2; 2, 5 respectively: their total degeneracies are 3×7; 2×5; 5×11, respectively.

Logical consistency might be better satisfied if the angular momenta of one-electron wavefunctions were represented in an analogous way. Certainly, we use the same letters to represent the *l* values—but in lower-case here to maintain our convention—which is why we talk of *s*, *p*, *d*, . . . orbitals. It is now clear, incidentally, why there is only one *s* orbital, but three *p*, five *d*, etc. But if we complete the lexigraphic analogy and write ^{2s+1}l, we find that the orbitals always lead with a left superscript of 2, because the *s* value for a single electron is always ½ and hence the spin-degeneracy is always 2. So the spin label is redundant and not written. However, because it is not written, our little discussion earlier about orbitals and spin-orbitals became necessary in a first-time view of the subject.

3.3 *Relationships between l and L; between s and S*

Most of what we have seen in this and the preceding chapter derives from the commutation properties of the quantum mechanical, angular momentum operators. Because of the form of the one-electron angular momentum eigenvectors that derive from those properties, we can similarly summarize another mathematical 'black box' by announcing that the same commutators are also responsible in the end for establishing the relationships between the upper and lower case angular momentum quantum numbers. It was emphasized above, on two separate occasions, that the eigenvalues of L^2, l^2, S^2, s^2 operators involve the product of 'a number times that number plus one'—the same rule in each case. But so far nothing has been stated or implied about the relationship between quantum numbers referring to the upper and lower case operators. The relationships are given, in fact, by what is variously known as the '*vector coupling rule*' or the '*vector triangle rule*'. The rule takes the same form regardless of the type of angular momentum—say, orbital or spin—being discussed. It is, that the quantum number describing the angular momentum of a pair of coupled particles may assume all integrally separated values ranging the sum to the difference of the quantum numbers labelling the individual particle motions: in symbols,

$$L \rightarrow l_1 + l_2, l_1 + l_2 - 1, \ldots |l_1 - l_2| \qquad (3.23)$$

$$S \rightarrow s_1 + s_2, s_1 + s_2 - 1, \ldots |s_1 - s_2|. \qquad (3.24)$$

As the word 'coupling' implies something between *two* entities, it should not be too surprising that the vector coupling rule refers to the interaction of only two particles at a time. For systems with more interacting particles, it is necessary to adopt a sequence of applications of the rule: couple l_1 with l_2 to form say l_{12}; then l_{12} with l_3 to give l_{123}; and so on until all particles have been accounted for. The process is lengthier but no new physical principles are involved.

It is useful to illustrate the vector coupling rule by a simple example and simultaneously account for the alternative name of the rule. We use a semi-classical analogue loosely and in the spirit of our discussion about coupled pendula. In Figure 3.3 electron orbital angular momenta of l_1 and l_2 units are represented by vector arrows and we know these quantities may be added vectorially to give a total angular momentum L for the pair. Quantization is grafted onto this classical description by requiring that only integral L values ranging the vectorial sum to difference of the two l values are allowed. The left-hand sketch shows the situation where both electrons classically rotate in the same sense and in parallel orbits: clearly, the total orbital angular momentum is given by their sum. On the right is shown the case in which the individual angular momenta are exactly opposed, in sense if not magnitude. The

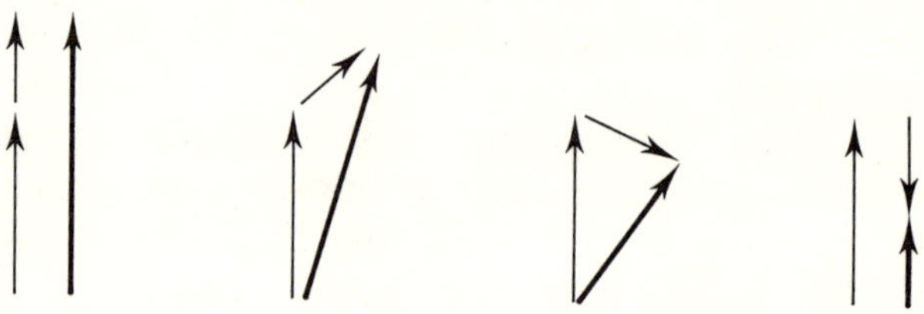

Figure 3.3 Vectorial addition of $\mathbf{l}_1$ and $\mathbf{l}_2$ to give $\mathbf{L}$: illustration of the vector triangle rule.

resultant is obviously given by the difference between the individual momentum vectors and, because we are ultimately representing total angular momenta squared here, we take the modulus to ensure a positive number. In between are vector diagrams representing intermediate (quantized) cases. The three vectors l_1, l_2, L must complete a triangle and that lies behind the alternate name—the vector triangle rule. Proofs are given elsewhere:[7,8] all we need here are the descriptions (3.23) and (3.24).

3.4 *Terms arising from the d^2 configuration*

We now learn a great deal by applying the vector coupling rule to the case of two *d*-electrons outside closed shells, provided we seek at all times to understand what lies behind the formal manipulations. Let us learn something of the properties of the eigensolutions for vanadium(III) ions—configuration $3d^2$. (By the way, we shall define 'configuration' more carefully in §3.5.)

The orbital angular momentum quantum numbers for *d* electrons are both 2 (Table 3.1), so $l_1 = l_2 = 2$. The spin quantum number for an individual electron is (always) ½, so $s_1 = s_2 = ½$. The vector triangle rule then predicts that possible L and S values are given by

$$\begin{aligned} L &\rightarrow 4, 3, 2, 1, 0 \\ S &\rightarrow 1, 0 \end{aligned} \qquad (3.25)$$

Note the word 'possible': we shall have more to say on this. The possible terms that might arise for the d^2 configuration follow on recognition of the independence of the space and spin properties, and we get the set

$$\begin{matrix} ^3G & ^3F & ^3D & ^3P & ^3S \\ ^1G & ^1F & ^1D & ^1P & ^1S \end{matrix} \qquad (3.26)$$

As the degeneracy of any given term ^{2S+1}L is $(2S + 1)(2L+1)$, we compute the total degeneracy of this set of terms as 27 + 21 + 15 + 9 + 3 + 9 + 7 + 5 + 3 + 1 = 100: a quicker way is to note that we have a spin triplet and singlet for each $L = 4$ to 0 value; that is, 4 × (9 + 7 + 5 + 3 + 1) = 100.

Now we look at the coupling of two *d*-electrons from a slightly different viewpoint. In equation (3.12) we established the commutation of L_z and H. The individual l_z operators, however, do not commute with H. Thus, taking the relevant parts of $l_z(1)$ and H we find the commutator is proportional to

$$\frac{\partial}{\partial\phi_1}(\phi_1 - \phi_2) - (\phi_1 - \phi_2)\frac{\partial}{\partial\phi_1} = 1 - (\phi_1 - \phi_2)\frac{\partial}{\partial\phi_1} \neq 0. \tag{3.27}$$

Altogether, the process of coupling (by Coulombic or other means) a pair of atomic electrons, replaces one set of commuting operators by another; namely

$$\left\{\begin{matrix} l^2(1)\ l_z(1)\ l^2(2)\ l_z(2) \\ s^2(1)\ s_z(1)\ s^2(2)\ s_z(2) \end{matrix}\right\} \rightarrow \left\{\begin{matrix} l^2(1)\ l^2(2)\ L^2\ L_z \\ s^2(1)\ s^2(2)\ S^2\ S_z \end{matrix}\right\}. \tag{3.28}$$

Correspondingly, we replace one set of 'good' quantum numbers by another

$$\left\{\begin{matrix} l_1\ m_{l1}\ l_2\ m_{l2} \\ s_1\ m_{s1}\ s_2\ m_{s2} \end{matrix}\right\} \rightarrow \left\{\begin{matrix} l_1\ l_2\ L\ M_L \\ s_1\ s_2\ S\ M_s \end{matrix}\right\}. \tag{3.29}$$

If any operator ceases to commute with the Hamiltonian—for example, $l_z(1)$ in the coupled case—eigensolutions of H fail to satisfy that operator also. Therefore, we cannot simultaneously measure definite values for the corresponding quantity and so eigenvalues and their labels—quantum numbers—cease to be meaningful. If, however, the unspecified coupling process is rather weak (corresponding to the weak spring in our pendulum example earlier), the commutators of the operators are small so that eigenfunctions of H almost satisfy the offending operator and the associated quantum member is approximately defined. Then we can use the jargon of a quantum number being 'good', 'fair' or 'bad', even though the notion of a continuum in the description of quantum numbers is strictly illogical from the beginning.

In the shift represented by either (3.28) or (3.29), note that the number of degrees of freedom is unchanged. This is part of a principle known as Ehrenfest's adiabatic law, the rest of which asserts that for a virtual, infinitely slow change in coupling strength, the quantum numbers of a system do not change. The philosophical import of this concept is less immediate today than it was when the old quantum theory was giving way to the new. We merely employ the general ideas to help introduce the idea of vanishingly small coupling. For a great deal of what follows, we play, in effect, a counting game. Let us return to the specifics.

We construct Table 3.2—which is only partial—to describe the possible sets of quantum numbers, comprising m_l and m_s values for each electron of the d^2 system. The presentation depicts the situation of vanishingly small coupling.

On the right is shown an alternative symbolism which is hopefully self-evident. A *microstate* represents formally uncoupled pairs of electrons labelled by their m_l values, with the signs of the m_s values ($= \pm\frac{1}{2}$) written above: the unwritten l and s values (here

Table 3.2

m_l	2	1	0	−1	−2	Microstate
	↑	↑				$(\overset{+}{2}\ \overset{+}{1})$
	↓		↓			$(\overset{-}{2}\ \overset{-}{0})$
	↑			↓		$(\overset{+}{2}\ -\overset{-}{1})$
	→ ↑↓					$(\overset{+}{2}\ \overset{-}{2})$
	~~↑↑~~					$(\overset{+}{2}\ \overset{+}{2})$
	~~↓↑~~					$(\overset{-}{2}\ \overset{+}{2})$
		↑		↓		$(\overset{+}{1}\ -\overset{+}{1}\ )$
			etc.			

2 and ½) are to be understood from the context. For a three-electron system, microstates are written in the form $(\overset{+}{2}\ \overset{-}{1}\ \overset{-}{0})$, for example; and so on for other electron configurations.

Table 3.2 is incomplete. It is a tedious task to complete it properly in this form but we can say at once how many entries it will contain. Each d electron may take any one of five m_l values and, at the same time either of two m_s values (α or β); making ten possibilities in all—$(2s+1)(2l+1)$ again. We wish to distribute two such electrons amongst these ten possibilities subject to two conditions. One is that we cannot distinguish the electrons—the labels 1 and 2 or 'first' and 'second' in the microstate bracket exist only in our imaginations—and the other is that no two electrons can share the same complete set of quantum numbers (by the Pauli exclusion principle which we shall discuss further in Chapter 5). Altogether, then, we have 10 possible descriptions of one electron times 9 for the second (Pauli), divided by two (electron indistinguishability); giving 45. More elegantly, we use the combinatorial formula,

$$^{10}C_2 = \frac{10!}{8!2!} = \frac{10.9}{2.1} = 45. \tag{3.30}$$

In Table 3.2, are shown examples of each kind of disallowed microstate where we have adopted a private convention to strike out once those microstates which are already given, but in a different, physically indistinguishable, order; and strike out twice those forbidden by the exclusion principle.

Now the 45 ways of placing two *d*-electrons within the atom—more correctly, of selecting angular momentum quantum numbers—describe 45 degrees of freedom for the d^2 system. But in (3.26) and following equations we used the vector coupling rule to establish a possible 100-fold degeneracy. Where is the inconsistency? Well there isn't one really: all is saved by the word 'possible' that was so pointedly emphasized earlier. Suppose, that instead of the *equivalent* electrons of the $3d^2$ configuration we have just examined, we had enquired about the *inequivalent* ones of $3d^14d^1$, for example. Use of the vector triangle rule would yield exactly the same result, of course, because $l_1 = l_2 = 2$ still. On the other hand, Table 3.2 would be replaced by one with two lots of five columns. There would be one electron in each set and problems arising from the Pauli exclusion principle and electron indistinguishability would not arise. The microstate $(\overset{+}{2}\ \overset{+}{2})$ in Table 3.2 would now be allowed, for each electron differs in its principal quantum number. Similarly the difference between $(\overset{+}{2}\ \overset{-}{2})$ and $(\overset{-}{2}\ \overset{+}{2})$ is more than the electron labelling sequence, for the first and second numbers carry with them implied information about the different principal quantum numbers again. Altogether, we now observe 10 possible m_l, m_s number pairs for the first electron and 10 for the second; total = 100 and the 'discrepancy' vanishes. Note, in passing, that while only ten microstates of the equivalent d^2 electron system are forbidden by the Pauli principle, 45 are disallowed by the general principle of indistinguishability of equivalent particles. In summary, then, about the use of the adjective 'possible' earlier, we observe that the vector coupling rule fails to take account of either indistinguishability or the Pauli principle and so can over-estimate the number of terms which can arise in any given case. The rule sets an upper limit on the variety of such terms.

The question arises, therefore, about which terms from the set (3.26) actually do arise for the equivalent d^2 configuration. In finding the answer we begin by recalling and emphasizing a few points covered earlier. First, the degeneracy of any term ${}^{2S+1}L$ is $(2S+1)(2L+1)$ always. Terms are necessarily complete: there is no physical meaning in a partial term with components other than exactly those summarized in Table 3.1. Secondly, using the vector triangle rule once more, we know that L for the d^2 system ranges 4 to 0. Therefore the maximum M_L value we shall encounter is +4 and the minimum is −4. Similarly, S ranges 1 to 0 and therefore M_S ranges +1 to −1. These M_L and M_S ranges allow us to set up Table 3.3

There are two aspects to this table—construction and use. In construction, the steps are as follows:

(1) Use the vector coupling rule to determine possible terms and hence establish the dimensions—that is, number of rows and columns—of the table by reference to the maximum ranges of M_L and M_S: this we have done.

Table 3.3

	M_S		
M_L	1	0	−1
4	~~$(\overset{+}{2}\ \overset{+}{2})$~~	~~$(\overset{+}{2}\ \bar{2})$~~	~~$(\bar{2}\ \bar{2})$~~
3	~~$(\overset{+}{2}\ \overset{+}{1})$~~	~~$(\overset{+}{2}\ \bar{1})$~~ ~~$(\bar{2}\ \overset{+}{1})$~~	·
2	~~$(\overset{+}{1}\ \overset{+}{1})$~~	~~$(\overset{+}{1}\ \bar{1})$~~	⋮
	~~$(\overset{+}{2}\ \overset{+}{0})$~~	~~$(\overset{+}{2}\ \bar{0})$~~ $(\bar{2}\ \overset{+}{0})$	
1	~~$(\overset{+}{1}\ \overset{+}{0})$~~	~~$(\overset{+}{1}\ \bar{0})$~~ ~~$(\bar{1}\ \overset{+}{0})$~~	⋮
	$(\bar{2}\ -\overset{+}{1})$	$(\overset{+}{2}\ -\overset{+}{1})$ $(\overset{+}{2}\ -\bar{1})$	
0	~~$(\overset{+}{0}\ \overset{+}{0})$~~	~~$(\overset{+}{0}\ \bar{0})$~~	⋮
	~~$(\overset{+}{1}\ -\overset{+}{1})$~~	~~$(\overset{+}{1}\ -\bar{1})$~~ $(\bar{1}\ -\overset{+}{1})$	
	$(\overset{+}{2}\ -\overset{+}{2})$	$(\overset{+}{2}\ -\bar{2})$ $(\bar{2}\ -\overset{+}{2})$	
−1			
−2	⋮	⋮	⋮
−3			
−4			

(2) Fill each location in the table with one or more microstates chosen so that $M_L = \Sigma_i^n m_l(i)$ and $M_S = \Sigma_i^n m_s(i)$, where n is the number of electrons being considered. The rationale behind this derives from Ehrenfest's adiabatic rule, for in the limit of vanishing coupling, the summations above are obviously valid. As we are engaged upon a sort of 'counting game', we can discover all we need to know at this stage by study of the limit in which the coupling process is deemed to have been gradually and endlessly weakened. Greater clarity will be brought to this aspect of the present procedure in §5.2. In the present case, then, the entry in the middle of the top row of the table arises because $2+2 = 4$ and $(+\tfrac{1}{2}) + (-\tfrac{1}{2}) = 0$. As the ordering of the electrons does not correspond to anything physical we do not write down $(\overset{-}{2}\ \overset{+}{2})$ as well, of course. The entry in the top row, column one—$(\overset{+}{2}\ \overset{+}{2})$—is disallowed through Pauli's exclusion principle, as we have discussed, but it and similarly forbidden microstates have been included explicitly in the table and then doubly struck out, for heuristic reasons. Table 3.3 is completed in like manner throughout. However, note the inherent symmetry that arises with respect to the signs of M_S and M_L values. As is repeatedly emphasized, we are merely counting—establishing a balance sheet—and so it suffices to know that the number of entries in columns to the right of the centre is the same as the number to the left; and that the number below a centre line is the same as above. So our table leaves most of the symmetry-related partners implicit.

(3) Check for completeness. In the left-hand column we count 4 allowed microstates above the middle row; this gives 8 after recognition of the symmetry-related microstates; add 2 for the case M_L=0 to get a total of 10 for the column M_S=1. The same is true for $M_S= -1$. Similar counting in the middle column gives $(2\times10)+5 = 25$; and a grand total of $10+10+25=45$. So all is well. This check is very important.

Now we *use* the table:

(4) Consider the list of possible terms to arise from the d^2 configuration, as given in (3.26), and begin with those with the greatest degeneracy. Here we shall consider them in the order ^3G, ^1G, ^3F, ^1F, ^3D, ^1D, ^3P, ^1P, ^3S, ^1S: other orders are possible and we comment upon this shortly.

(5) The term under consideration is $(2S+1)(2L+1)$-fold degenerate. First we have ^3G, for which $S = 1$ and $L = 4$. Therefore, M_S

values range 1, 0, −1 while M_L values cover 4, 3, 2 ... −3, −4. If a ^3G is to be possible—and it must be complete remember—we require one entry in each row of each column of the table. As there is no allowed microstate for $M_L=4$, $M_S=1$, a complete ^3G term cannot be constructed and so does not arise in the d^2 configuration. We turn to the case of ^1G for which $S = 0$ while $L = 4$. Here we require one entry in each row of the centre column only, for $S = 0$ implies $M_S = 0$ only. This *can* be done and we strike out the microstates accordingly. It does not matter which microstate in any table box is struck out, for at this stage (once more) we are merely engaged in a balance sheet exercise. We show here what has happened by striking out microstates with the lines which slope upwards to the right. Now, before moving on to the ^3F term, we consider the possibility that there might be more ^1G terms, for there is no objection to repeated terms arising from any given configuration: after all, they would still satisfy the vector coupling condition. However, in this case we note that we have run out of entries in the top middle box and so there is only one ^1G term. For the ^3F term, entries in all three columns of all rows from $M_L=3$ to −3, inclusive, are required. This also can be satisfied and the microstates have been struck out this time by lines which slope downwards to the right. Check for repeated terms—there are none actually for the d^2 case at all, but there are for d^3. Continuing in this way eventually produces the term list

$$^1G \quad ^3F \quad ^1D \quad ^3P \quad ^1S, \tag{3.31}$$

at which point the table has been emptied.

(6) Double check the procedure by computing the total degeneracy of the terms just produced: 9+21+5+9+1 = 45, as required.

Once the reader has worked through this example it should be clear why we consider possible terms in the sort of order given in (4) above. Suppose we had begun with ^1S. We strike out one of the five microstates in the centre column of the table. Repeats? yes, five of them! Then consider, say ^3S. This is impossible for it requires entries in all three columns of the middle row and none remain in the central column. In fact, we would find that no set of terms can be selected in this way that account for all the microstates once and only once. The reader may care to check and practice the procedure, however, by arriving at the correct, unique answer (3.31) by considering terms in the order ^3G, ^3F, ^3D, ^3P, ^3S, ^1G, ^1F, ^1D, ^1P, ^1S. Finally at this point we note how this process, though simple to understand, is actually rather clumsy, for consider what is required to do the same thing for the d^3 system, for example. Here there are 120 microstates ($= {}^{10}C_3$) of the form $(\overset{+}{2}-\bar{1}-\bar{2})$ which have to be placed in such a table and then systematically accounted for. There are, in fact, sophisticated methods which allow the possible terms of a configuration to be determined quickly and easily: unfortunately, a great deal of mathematical theory must be mastered first! So we will not bother. Instead we turn our attention to the meaning of all we have achieved so far in this chapter.

3.5 *Configurations, terms and Hund's first two rules*

In taking stock let us recall the original purpose of studying many-electron systems. We centre our discussion around Figure 3.4. The Hamiltonian for a many-electron atom involves $H_{\text{H-like}}$ for each electron plus a Coulomb operator which relates to the electrostatic interelectron potential energy. As we discussed in Chapter 1, the solutions to the one-electron Hamiltonian alone—that we now call orbitals—are quite distinct from those for the many-electron system—so-called term wavefunctions. Of course, the difference between orbitals and term wavefunctions involves radial factors as well as the angular parts we have discussed but we have not prepared the background to describe these. If we notionally separate the one- and two-electron operators in an atomic (or molecular) Hamiltonian as in Figure 3.4 and consider the individual electrons, therefore, to behave in ignorance of one another, we have an entity called an electronic *configuration*. In writing d^2, for example, we recognize two electrons possessing kinetic energy and being attracted to the (shielded) nucleus; each electron undergoes motion whose angular properties define the letter d, as we have seen. There are 45 equally good—that is, equi-energetic—ways of constraining two electrons while satisfying these requirements.

On the right of the diagram, we recognize that the electronic motions are coupled because of the electrostatic interaction accounted for by the Coulomb operator $1/r_{12}$. In the present example, we now observe that there are 21 equally best ways of arranging two electrons which have kinetic energy, are attracted to

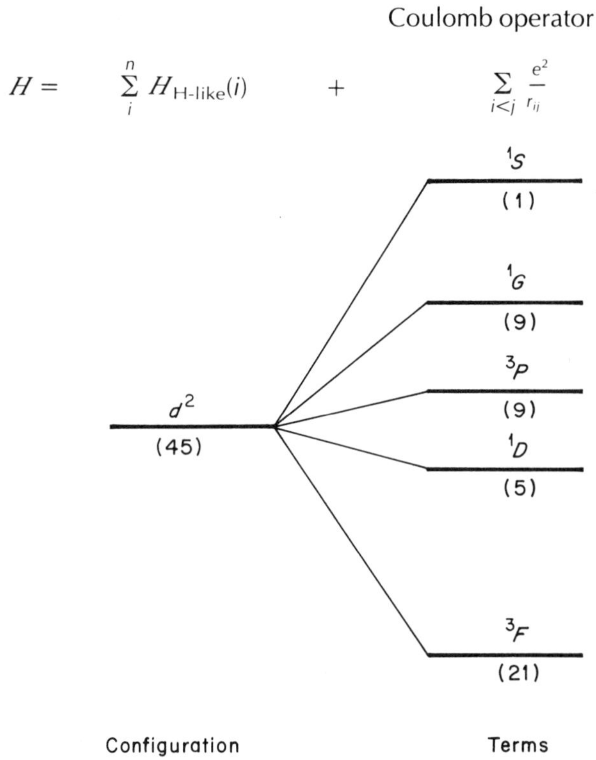

Figure 3.4 Terms arising from the d^2 configuration.

the nucleus, undergo motion with orbital angular momentum squared characterized by $L=2$, *and* which keep out of each other's way. The extra Coulombic requirement sets a more exacting group of conditions, therefore, which can be satisfied in a number of different kinds of way. Each way (term) is generally degenerate but, since the total number of electronic arrangements must remain unchanged, the number of possibilities associated with any one manner of coupling is necessarily less than the number of ways of satisfying the one-electron Hamiltonian alone.

As already mentioned, the term wavefunctions involve both angular and radial parts, so that the labels ^3F, ^1D, etc.—like the configuration label, d^2—only describe the angular parts. The energies of those two-electron functions are determined by solution of the appropriate Hamiltonian operator: the theory of angular momentum that we have studied offers no insight into energies. There is no simple method of computing these energies either, nor of determining their relative ordering. There do exist rules, however, by which the ground, *and only the ground*, term may be selected.

Hund's first rule: Of the terms selected by the application of angular momentum theory, the ground term will be one with maximal spin-degeneracy.
Hund's second rule: If ambiguity remains after application of the first rule, that term of maximum orbital-degeneracy lies lowest in energy.

In summary: maximum *S*, *then*, maximum *L*. In the present case, therefore, Hund's rules give the ^3F term as the ground term. To repeat: no rules exist which provide information on the absolute or relative energy of any other term. So far as our present discussion is concerned, the remaining four terms of the d^2 configuration have been arranged arbitrarily.

Let us put the angular momentum side of Figure 3.4 into a different perspective by emphasizing the primal cause of the splitting; namely, the Coulomb term in the Hamiltonian. A schematic aid, which must not be taken literally, goes as follows. In Figure 3.5 are sketched three different arrangements of two electrons around a nucleus characterized by progressively decreasing interelectron distances and increasing interelectron repulsions. The arrangements describe, therefore, from left to right, increasing

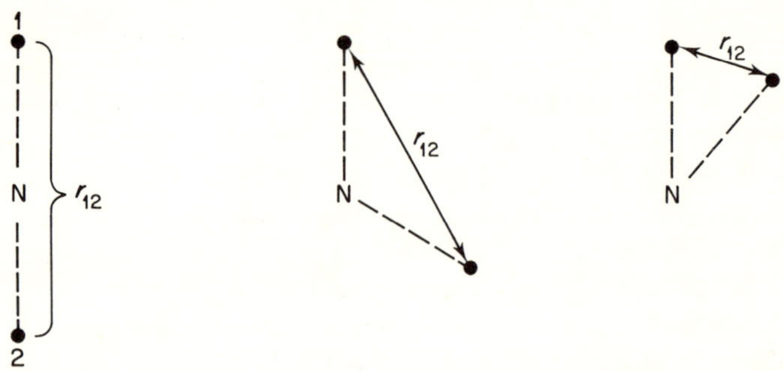

Figure 3.5 A pictorial mnemonic illustrating how different interelectron repulsion energies might be associated with different total orbital angular momenta.

energy conditions. We now imagine that each electronic arrangement is held fixed but may rotate around the nucleus. In this way, we can arrive at the viewpoint that the angular momentum properties of the various electronic arrangements are a consequence rather than a predictor of the different energy levels. Now this is not really so, for the properties of energy and angular momentum are inextricably linked in real atoms, but our picture might hopefully redress the balance a little away from an angular momentum-centred viewpoint. Following this approach a shade further: suppose we had tackled the problem of solving the many-electron Schrödinger equation head on. After a great deal of numerical computation, we would have found that 21 eigenfunctions, which later checked out to share common orbital and spin angular momentum, had equally minimum energies; that 5 eigenfunctions, sharing equal but different angular momentum characteristics, had equal second-best (say) energies; and so on. The procedures of angular momentum theory that we have studied in this chapter, once understood, provide half of this information without any need for a computer. Of course, the other half—the energies—is not accessible by qualitative means. The process *we* have examined refers to the so-called *vector coupling model* of atomic structure.

3.6 *Ground terms quickly*

Although the vector coupling methods are simple in principle, they can be quite lengthy, nevertheless: the d^3 configuration is 120-fold degenerate, for example, and gives rise to the term set; ^4F, ^4P, ^2H, ^2G, ^2F, ^2D plus two ^2P. For many purposes, it is sufficient to know only the ground term arising for any configuration. In these cases, there exists a trivally simple procedure for its determination:

(1) Draw $2l+1$ boxes: for example, for a d-electron problem, we require 5

(2) Label the boxes with m_l values decreasing algebraically from left to right:

m_l 2	1	0	-1	-2

(3) Since any ground term is of maximum spin multiplicity (Hund's first rule), electrons should be placed in these boxes in such a way as to maximize the number of unpaired and parallel spins. Also, since we require a maximum L value commensurate with this (Hund's second rule), we fill up the boxes from the left. For example, for d^2:

m_l	2	1	0	−1	−2
	↑	↑			

(4) Sum the electronic m_l values to get L and the spins to get S. Here; $L = 2+1=3$, i.e. an F term, and $S = ½+½=1$, whence the ground term ^3F. The rationale behind these additions is as follows. Had the electrons been placed in any other boxes, the sum of m_l values spanned would have ranged L to $-L$: in the present case 3 to -3, so implying $L = 3$.

Just one more example. For the d^8 configuration, we have

m_l	2	1	0	−1	−2
	↑↓	↑↓	↑↓	↑	↑

and $\Sigma\, m_l = (2+-2) + (1+-1) + (0+0) + 2 + 1 = 3$ and $\Sigma\, m_s = 5 \times (½) + 3\times(-½) = 1$ giving the ground term ^3F. Remember: this quick method only works for the ground term—see if you can understand why. Also, to gain familiarity, why not determine ground terms arising from the following configurations:

(a) d^1 to d^9
(b) f^2, f^3, f^6, f^8, f^{12}, f^{13}
(c) s^0, s^1, s^2 / p^0, p^3, p^6 / d^0, d^5, d^{10} / f^0, f^7, f^{14}.

Observe any patterns that you find in the answers.

CHAPTER 4

Spin-Orbit Coupling

4.1 *Electrostatic versus magnetic coupling*

The interelectronic coupling we have studied so far arises from the electrostatic interaction between electrons and is represented by the Coulomb operator e^2/r_{12}. Coupling can also occur, however, by an entirely separate mechanism. This is magnetic in origin and the phenomenon is called spin-orbit coupling.

The magnetic properties of atoms are intimately connected with angular momentum. Classically, we view an electron orbiting a nucleus as a small electric current loop which, as we know from school physics, generates a magnetic field perpendicular to the plane of the loop. Just as we drawn an arrow in Figure 3.3 to represent an electronic angular momentum, we can utilize that same arrow to represent the attendant magnetic moment generated by the orbiting electron. It does not matter that the path of an electron is not described in the quantum mechanical picture for we still deal with observable quantities associated with the *average* motion. But even in a one-electron atom, the electronic orbital motion is not the only source of internal magnetic moment for the electron possesses an intrinsic spin. With this form of angular momentum is associated another magnetic moment which, for reference purposes, we call the spin magnetic moment. For some it may be helpful to think of the electron as a point charge rotating on some axis and so generating a magnetic field. Classically, no such field can be produced in this way from a point source. If we relax the size condition and view the electron as having a non-vanishing radius, classified physics would require an associated magnetic field although—and like the earlier orbiting motion—the particle should radiate energy and collapse inwards. Quantum theory rescues us from these problems, however, as we know. Furthermore, since spin is a relativistic effect, as discussed in §2.8, we cannot expect everyday explanations of it to be totally convincing: for that matter, magnetic fields are relativistic too. So from now on we take as given, that atomic electrons possess (intrinsic) spin magnetic moments and (extrinsic) orbital magnetic moments. To zeroth order in Dirac's electronic spin theory, these properties are independent of one another.

In first order, however, they are not. What we call spin-orbit coupling is not so much a separate and distinct physical phenomenon as an additional term in the relativistic spin theory which arises when we implement that theory in more detail. We represent this first-order correction term by the one-electron spin-orbit coupling operator,

$$H_{\text{SOC}} \equiv \zeta \, \mathbf{l}.\mathbf{s}. \tag{4.1}$$

Here $\mathbf{l}$ and $\mathbf{s}$ are the orbital and spin angular momentum operators (not squared) for the electron, and ζ is a quantity, with energy units, called the *one-electron spin-orbit coupling coefficient.* Note straightaway that the dimensionality of the spin-orbit operator is appropriate. The point here is that the radial part must arise in an operator which is part of a Hamiltonian, even though the coupling process only involves angular momentum operators which do not refer to the radial coordinate. The spin-orbit coupling coefficient may be thought of as a scaling parameter. In fact the symbol ζ subsumes a radial integral and so the operator (4.1) strictly acts only upon the angular coordinates of wavefunctions.

A classical representation of spin-orbit coupling can be helpful and we refer to Figure 4.1 for the first of several times in this chapter. An electron is shown there as moving around the nucleus in a Bohr-like orbit. Being (negatively) charged, the motion is

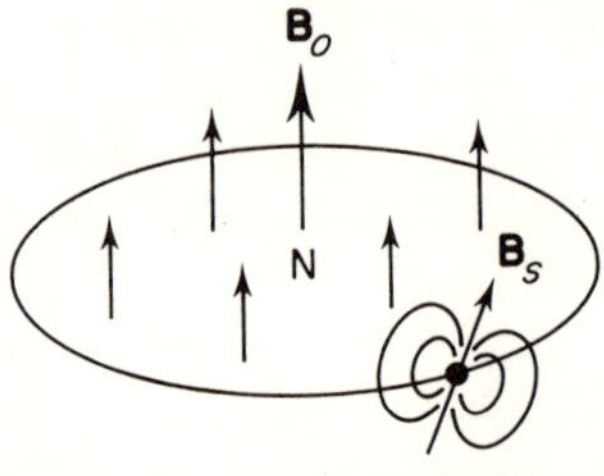

Figure 4.1 A classical view of spin-orbit coupling: the magnetic field of the electron spin, $\mathbf{B}_s$, interacts with that produced by the orbital motion, $\mathbf{B}_o$.

equivalent to an electric current and B_o represents the magnetic field produced by the orbital motion. Meanwhile the electron spins on its own axis, or possesses a spin magnetic moment, and generates the field B_s. Clearly the spin and orbital motions of the electron will tend to orient themselves in such a way as to minimize their interaction energy as mediated through these magnetic fields: and the effect is, of course, mutual. Recalling our pendulum analogue in §3.1, we conclude that the motion of the electron around the nucleus will not be characterized by a constant angular velocity—there will be precessional character of the motion in fact—and so under the present conditions of magnetic coupling, m_l and m_s will no longer be good quantum numbers.

Just as in equations (3.1) to (3.8), where the electrostatic interaction obliged us to define new total orbital and spin angular momentum operators, the magnetic coupling now requires the introduction of a total angular momentum operator. But note that this is not, at this stage, anything to do with interaction between *different* electrons: we are concerned with the coupling of two internal degrees of freedom for any one given electron. We write a total angular momentum operator for a single electron as

$$\mathbf{j} = \mathbf{l} + \mathbf{s} \tag{4.2}$$

and analogous to (2.15), (2.16), (2.35) and (2.36), we have

$$\hat{j}^2\psi = \{\hbar^2\, j(j+1)\}\,\psi \tag{4.3}$$

$$\hat{j}_z\psi = \{\hbar\, m_j\}\psi \tag{4.4}$$

where j takes integrally separated integer or half-integer values: $j \rightarrow 0, 1, 2 \ldots$ or $j \rightarrow \frac{1}{2}, \frac{3}{2}, \frac{5}{2} \ldots$; and m_j ranges j to $-j$ in integral steps. Like (3.28) and (3.29), we follow the coupling process by noting the changes in the set of commuting operators and their associated eigenvalues:

$$\{\hat{l}^2\ \hat{s}^2\ \hat{l}_z\ \hat{s}_z\} \quad \rightarrow \quad \{\hat{l}^2\ \hat{s}^2\ \hat{j}^2\ \hat{j}_z\} \tag{4.5}$$

$$\{l\ \ s\ \ m_l\ m_s\} \quad \rightarrow \quad \{l\ \ s\ \ j\ \ m_j\} \tag{4.6}$$

independent spin and space functions	magnetic, spin-orbit coupled momenta

While l and s are referred to as quantum numbers labelling orbital and spin angular momentum, j is said to label total angular momentum.

The vector coupling rule that derives from the commutation relationships of angular momentum, which are also satisfied by j^2 and j_z, gives values for j according to

$$j \rightarrow l + s,\ l + s - 1, \ldots .\ |l - s|\ , \tag{4.7}$$

a rationale for which comes readily by reference to Figure 3.3 and associated remarks. As asserted in §2.2, the consequences of angular momentum and commutation are amazingly pervasive.

4.2 *Electrostatic and magnetic coupling together*

Once we consider atoms with more than one valence electron, the situation arises in which electronic motions are coupled together by what are, in the present exposition, two separate physical causes. As any coupling process, by its very name, involves just two entities at a time, a question of precedence must now be considered.

The problem is illustrated for the two-electron case by Figure 4.2. The key to an understanding of this issue is to recognize that spin-, orbital- or total angular momenta are all quantities of the same kind. Indeed, because spin-orbit coupling can be regarded as an artefact of the stepwise, Schrödinger-to-Dirac, description of the system, all three may be viewed as different aspects of the very same quantity. Working horizontally in the top two rows of the diagram are represented the coupling processes arising out of the Coulomb interaction, as described in the preceding chapter. Moving down the first two rows we see the effect of spin-orbit coupling on each electron (1 and 2) to produce total angular momentum numbers for each—j_1 and j_2. Across the bottom line is represented the coupling together of $\mathbf{j}_1$ and $\mathbf{j}_2$ by the Coulombic interelectron mechanism. Here we summarize the change in commuting operator set and changes in good quantum numbers:

Operators:

$$\left\{\begin{matrix} l^2(1) & s^2(1) & j^2(1) & j_z(1) \\ l^2(2) & s^2(2) & j^2(2) & j_z(2) \end{matrix}\right\} \rightarrow \left\{\begin{matrix} l^2(1) & s^2(1) & j^2(1) & l^2(2) \\ s^2(2) & j^2(2) & J^2 & J_z \end{matrix}\right\} \tag{4.8}$$

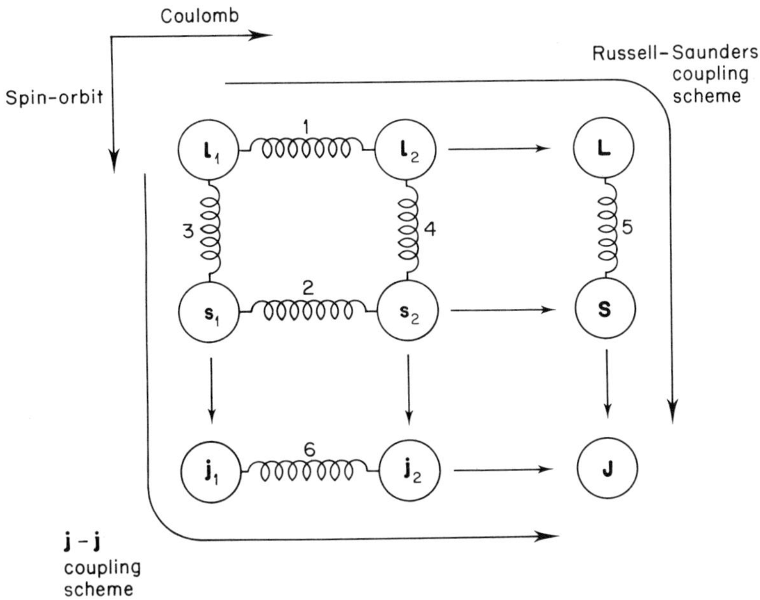

Figure 4.2 Different coupling sequences define the Russell–Saunders and **j–j** coupling schemes.

Quantum numbers:

$$\begin{Bmatrix} l_1 & s_1 & j_1 & m_{j1} \\ l_2 & s_2 & j_2 & m_{j2} \end{Bmatrix} \rightarrow \begin{Bmatrix} l_1 & s_1 & j_1 & l_2 \\ s_2 & j_2 & J & M_J \end{Bmatrix} \quad (4.9)$$

Independent electrons coupled orbit/spin by s.o.c. — Electron pair after Coulombic coupling

Once more, the vector coupling rule provides values for the total J which, being a label for the electrons as a group, is written in upper case:

$$J \rightarrow j_1 + j_2, j_1 + j_2 - 1, \ldots |j_1 - j_2| . \quad (4.10)$$

The eigenfunctions for the electron set, coupled by both electrostatic and magnetic means, satisfy the equations:

$$J^2 \mid J, M_J\rangle \quad = \{\hbar^2 J(J+1)\} \mid J, M_J\rangle \quad (4.11)$$

$$J_z \mid J, M_J\rangle \qquad = \{\hbar M_J\} \mid J, M_J\rangle \,. \tag{4.12}$$

Returning to the diagram; the vertical transformation on the right-hand side represents the reverse order of coupling in which the spin and orbital angular momentum properties of the group (here, pair) of electrons are magnetically coupled by the spin-orbit operator. Instead of

$$\mathbf{J} = \mathbf{j}_1 + \mathbf{j}_2 \,, \tag{4.13}$$

like (4.2), for the case just described, we write

$$\mathbf{J} = \mathbf{L} + \mathbf{S} \tag{4.14}$$

and monitor the changes in commuting operators and associated quantum numbers by

Operators:

$$\begin{Bmatrix} l^2(1) & l^2(2) & s^2(1) & s^2(2) \\ L^2 & S^2 & L_z & S_z \end{Bmatrix} \rightarrow \begin{Bmatrix} l^2(1) & l^2(2) & s^2(1) & s^2(2) \\ L^2 & S^2 & J^2 & J_z \end{Bmatrix} \tag{4.15}$$

Quantum numbers:

$$\begin{Bmatrix} l_1 & l_2 & s_1 & s_2 \\ L & S & M_L & M_s \end{Bmatrix} \rightarrow \begin{Bmatrix} l_1 & l_2 & s_1 & s_2 \\ L & S & J & M_J \end{Bmatrix} \tag{4.16}$$

Two electrons coupled by the electrostatic Coulomb interaction	The same after further coupling by the spin-orbit operator

Yet again, the vector triangle gives values for J in terms of its components:

$$J \rightarrow L + S,\, L + S - 1, \ldots \,|L - S| \,. \tag{4.17}$$

Of course, once both processes have done their work, the final states are still eigenfunctions of J^2 and J_z, as in (4.11) and (4.12). The eigenfunctions are not the same in other respects, however, in that the first set discussed here are also eigenfunctions of $j^2(1)$ and $j^2(2)$ while the second set are eigenfunctions of L and S instead. If the reader is a little puzzled here about how the, apparently, same result (bottom right of Figure 4.2) *must* be described in different ways, let him be assured that all is well and this point will be clarified later on.

4.3 *The Russell–Saunders coupling scheme*

We have seen that a description of angular momenta in atoms can yield total angular momenta for the system characterized by the quantum numbers J, M_J, by two different notional routes. Which route is the more appropriate depends upon which of the two intermediate situations—characterized either by L and S or by j_1 and j_2—more closely approximates the final situation. If, for example, the electrostatic coupling process is very much greater than the magnetic, an incomplete coupling analysis yielding the terms described in the preceding chapter, will provide a reasonably acceptable picture of atomic energy levels. Subsequent 'refinement' by spin-orbit coupling is then based on terms as a starting point. In fact this order of coupling, which is known as the Russell–Saunders coupling scheme, is by far the more appropriate throughout the periodic table. The reverse order of interaction—magnetic followed by electrostatic—is called the **j–j** coupling scheme. We discuss it in §4.8. Referring to Figure 4.2 again, the Russell–Saunders scheme describes the situation corresponding to the electrostatic coupling 'springs', 1 and 2, being stiff while the magnetic ones, 3 and 4, are weak. The two pairs of electrons labelled by l_1/l_2 and s_1/s_2 will execute motions in such discrete pairs at first glance. Only after a prolonged inspection would we discern the action of the magnetic coupling spring, 5, causing the frequencies of these pairs to vary slowly. Roughly, therefore, a description in terms of L and S as 'fairly good' quantum numbers would be a reasonable approximation to a more correct description in terms of the totally coupled momenta labelled by J and M_J.

Sets of wavefunctions deriving from a Russell–Saunders term, ${}^{2S+1}L$, which share a common J value are called *levels* and labelled by the symbol

$$ {}^{2S+1}L_J \tag{4.18} $$

and, because the M_J values associated with any given J value range J to $-J$ in integral steps, the total degeneracy of a level is $(2J+1)$. We use the word 'terms' for many-electron functions notionally produced before spin-orbit coupling; and 'levels' for all functions produced after spin-orbit coupling.

Now consider ^3F arising as the ground term of the configuration d^2. The term label tells us that $L = 3$ and $S = 1$; so, using the vector triangle rule (4.17), we have $J = 4, 3, 2$. Spin-orbit coupling thus splits up the degeneracy of the ^3F term into three levels, $^3F_4$, $^3F_3$ and $^3F_2$ which are 9-, 7-, and 5-fold degenerate, respectively. Note once more, how the number of degrees of freedom—the number of electronic arrangements within the term—remains unchanged: $9 + 7 + 5 = 21$, the $(2S + 1)(2L + 1)$ degeneracy of the parent ^3F term. Notice also, that the use of the vector coupling rule now provides a *complete* description of the species produced in the coupling process, for the problems of Pauli's exclusion principle and the indistinguishability of electrons have been dealt with already.

4.4 *Hund's third rule and Landé's interval rule*

The group of levels deriving in this way from a term is called a *term multiplet*. The relative *energies* of the levels belonging to a term multiplet are not determined completely by the rules of angular momentum and vector coupling theory, of course, for these deal only with angular coordinates. As usual, it is the Hamiltonian operator, here given by

$$H_{\text{soc}} = \lambda\, \mathbf{L.S}, \tag{4.19}$$

that finally establishes level energies. This operator is the many-electron equivalent of $\zeta \boldsymbol{l.s}$ used for the one-electron process. The quantity λ is called the *many-electron spin-orbit coupling parameter*: once again, it subsumes integrations over the radial coordinate and has units of energy. The same value of λ applies for all terms arising from the same configuration and the many- and one-electron coefficients are related by the equation

$$\lambda = \pm \frac{\zeta}{2S}. \tag{4.20}$$

The single-electron coefficient ζ is an intrinsically positive quantity while (4.20) shows that λ may be positive or negative. Which sign λ

takes is determined at this stage, by assertion; indeed by a rule known as *Hund's third rule*. At this point in the story, we state the rule in a non-standard, though perfectly correct, way.

Hund's third rule: For atoms with less than half-filled l^n configurations, λ is positive: for atoms with more than half-filled shells, λ is negative. For example: λ is positive for d^2 and d^3; negative for d^7 and d^8. We discuss the case of exactly half-filled shells later.

While Hund's third rule establishes the *absolute* ordering of level energies arising from any one given term, the *relative splittings* are determined by the *Landé interval rule*. In words, this states that the energy separation between two levels arising from the same term and characterized by integrally adjacent J values is λ times the larger of the two J values.

Proof of Landé's interval rule

Consider two levels possessing angular momentum labelled by L, S, $J+1$ and L, S, J, respectively, arising from a term ^{2S+1}L, as in Figure 4.3. We seek an expression for the energy separation ΔE between them.

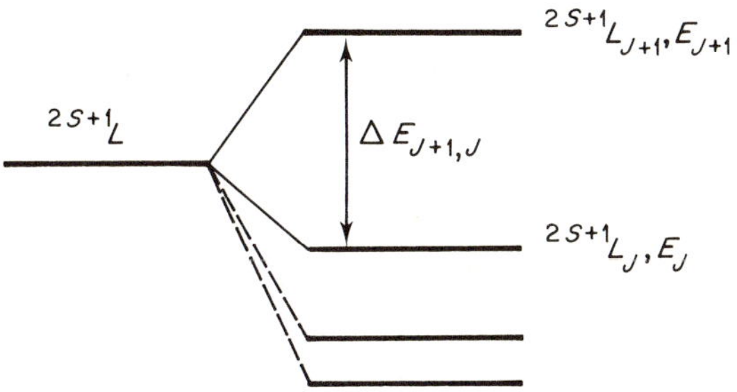

Figure 4.3 Two levels arising from the same term.

Now the total angular momentum operator J is defined by (4.14) and the many-electron spin-orbit coupling operator by (4.19). Therefore, since

$$\mathbf{J} = \mathbf{L} + \mathbf{S}, \tag{4.14}$$

$$J^2 = L^2 + S^2 + 2\mathbf{L.S}$$

whence

$$\lambda\mathbf{L.S} = \frac{\lambda}{2}(J^2 - L^2 - S^2)\,. \tag{4.15}$$

While we had not previously established rules by which to manipulate the product **L.S**, (4.15) now facilitates this by way of (3.16), (3.18) and (4.11). Since the energies of orthonormal wavefunctions are related to the Hamiltonian operator by

$$E = \langle\psi \mid \mathcal{H} \mid \psi\rangle$$

we then find

$$\begin{aligned} E(^{2S+1}L_J) &= \langle LSJ \left| \frac{\lambda}{2}(\hat{J}^2 - \hat{L}^2 - \hat{S}^2) \right| LSJ\rangle \\ &= \hbar^2 \frac{\lambda}{2} \langle LSJ \mid J(J+1) - L(L+1) - S(S+1) \mid LSJ\rangle \\ &= \hbar^2 \frac{\lambda}{2} \{J(J+1) - L(L+1) - S(S+1)\}\,. \end{aligned} \tag{4.16}$$

The second line follows because functions characterized by L, S, and J are *eigen*functions of $\hat{J}^2$, $\hat{L}^2$ and $\hat{S}^2$ and so, by (4.11), (3.16) and (3.18), are all merely multiplied by the numbers $J(J+1)\hbar^2$, etc. The third line follows on recognition that bra and ket in $\langle LSJ|LSJ\rangle$ are the same.

The energy of the level $^{2S+1}L_{J+1}$ may be calculated similarly; but more simply by substituting $J+1$ for J in (4.16):

$$E(^{2S+1}L_{J+1}) = \hbar^2 \frac{\lambda}{2} \{(J+1)(J+2) - L(L+1) - S(S+1)\} \tag{4.17}$$

Subtracting (4.16) from (4.17) gives

$$\Delta E_{J+1,J} = \hbar^2 \lambda\, (J+1) \tag{4.18}$$

or, since it is conventional to work in atomic units in this area, when $\hbar \rightarrow 1$ we get the more usual form of Landé's rule,

$$\Delta E_{J+1,J} = \lambda\,(J+1). \tag{4.19}$$

Note that the origin of this level energy separation involving only J and not L or S lies in the common L and S values shared by levels from the *same* term as resulting in the cancellation of the $L(L+1)$ and $S(S+1)$ parts of the subtraction of (4.16) and (4.17).

A trivial extension of Landé's interval rule is that the complete multiplet of levels arising from one given term are ordered energetically in the sequence of their J values. So for the ^3F term of the d^2 configuration, for example, we find the splitting pattern given in Figure 4.4. Further, since the configuration d^2 comprises a less than half-filled shell, Hund's third rule, establishes the absolute ordering shown, as λ is positive. A more common form of Hund's third rule may now be stated.

Figure 4.4 The levels arising from the ground term of the d^2 configuration.

Hund's third rule. In atoms with *less* than half-filled l^n configurations, that level with *minimum* J arising from a given term lies lowest energetically: conversely, for atoms with *more* than half-filled shells, the level with *maximum* J lies lowest.

In the former case we refer to a *normal multiplet* and when λ is negative—and the levels appear in reverse order—to an *inverted multiplet*.

4.5 *Hole formalisms*

There are some close similarities between d^n and d^{10-n} configurations, for example. Consider the d^1 and d^9 cases. The total degeneracy of the one-electron case is, ${}^{10}C_1 = 10!/9!1! = 10$, while that of the d^9 configuration is ${}^{10}C_9 = 10!/1!9! = 10$. From the point of view of the combinatorial formula, arranging one object amongst ten boxes is precisely the same as leaving one vacancy in a full set of ten boxes. In short, d^9 may be thought of as one hole in a full d shell. Similar arguments relate p^n and p^{6-n}; f^n and f^{14-n}, etc. Physically, however, there are two differences between d^1 and d^9 that must be recognized. One is that the hole constitutes a positive entity in a full 'sea of electrons' rather than a negatively charged object. The other is that screening and other effects relating to the total electron density are not mirrored between 'electron-hole' pairs so that observables associated with absolute energies are not simply related. But angular momentum properties *are* very closely related and display special sign properties.

If we view the d^9 configuration as a distribution of one hole in the d shell, then spin-orbit coupling effects will be characterized by a sign change with respect to the d^1 configuration. This follows from the reversed charge sign of the orbiting particle. Generally, any one-electron operator will behave in this way with respect to 'reciprocal configurations': another example, is the ligand-field operator referred to as the ligand-field potential. The same sign change is responsible for the reversal in ground state predicted by Hund's third rule as one traverses the half-filled shell.

Although we are primarily concerned with spin-orbit coupling in this chapter, it is sensible to include here a comment about the hole formalism associated with the two-electron operator, e^2/r_{12}. The reader who worked through the problems in §3.6, will have discovered that the ground terms of 'reciprocal configurations' are identical. Actually, *all* terms arising from such configuration pairs are identical. The reason why they share common *ground* terms, however, rather than there being an inversion at the half-period, is because the two holes in d^8, say, *repel* one another in exactly the same qualitative way as the two electrons *repel* one another in d^2. So for the *two*-electron Coulomb operator, we do not see an inversion. However, the absolute energies of the terms (${}^3F$, ${}^3P$, ${}^1D$, ${}^1G$, ${}^1S$) arising from d^8 differ from those arising from d^2: indeed the relative ordering can differ also. Again, this occurs because the 'mirror' behaviour does not extend to the quantitative aspects of these systems, for shielding and so on are not properties related in so simple a manner.

4.6 *Russell–Saunders coupling review: configurations, terms, levels*

We now summarize and exemplify what we have learned so far about the Russell–Saunders coupling scheme. Following Figure 3.4, we consider successive perturbations upon the d^2 configuration but now by both electrostatic and magnetic processes, as in Figure 4.5.

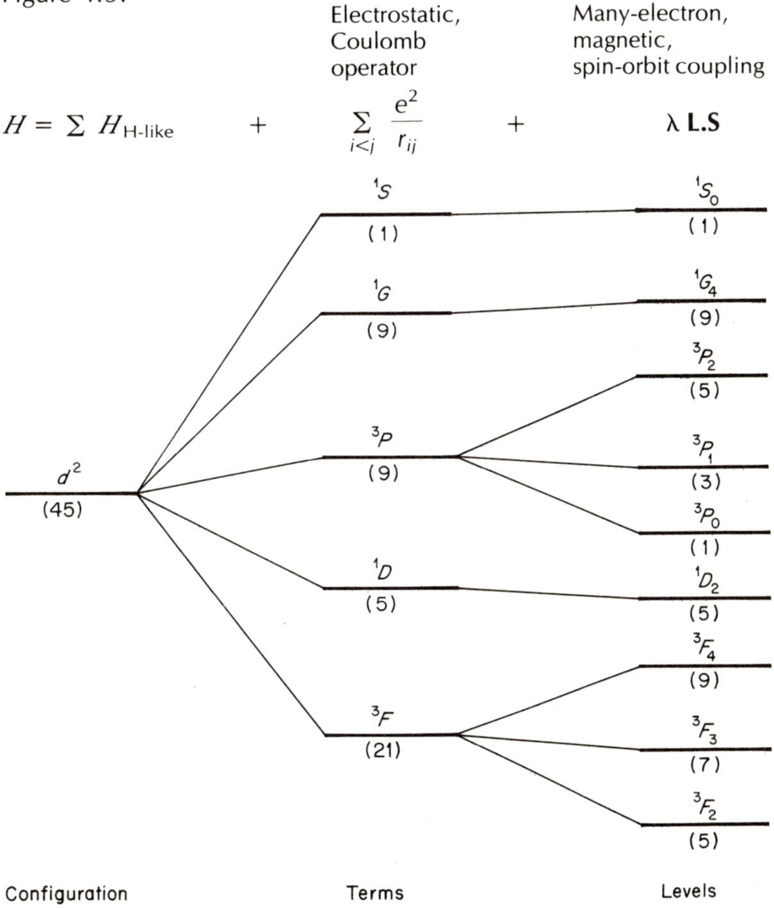

Figure 4.5 Splitting of the complete d^2 configuration in the Russell–Saunders coupling scheme.

For those atoms in which electrostatic coupling dominates the magnetic (which are the majority as we shall see), the RS scheme sensibly deals with these effects in this order, the level energy splittings on the right being much smaller (say, by one-to-two orders of magnitude) than the term splittings in the middle. For example, for V^{3+} ions, d^2, and ${}^3F$–${}^1D$ separation is of the order 20 000 cm^{-1} while the multiplet width of ${}^3F$, corresponding to 7λ, is about 700 cm^{-1}. Again, by way of summary: if we consider only those levels which arise from any one term, the multiplet width or splitting pattern are given by Landé's interval rule. The reason for the proviso here will become clear shortly. Suffice it to say that the quantitative splitting descriptions are strictly correct only in first order.

Problems

(a) Determine all first-order details of the level splittings for the ground terms of d^3, d^7, f^6 and f^8 configurations.

(b) An excited term arising from the d^5 configuration is ${}^4P$. How does it split under spin-orbit coupling? Problems raised by this example, form part of the ensuing text.

4.7 *Magnitudes of spin-orbit coupling coefficients*

The spin-orbit ζ and λ values vary widely throughout the periodic table and since the relative size of the spin-orbit and Coulomb perturbations is an important issue, we look now at how large these variations are and why they occur.

Once more, recall the relativistic origins of the spin-orbit coupling mechanism. An explanation with any degree of rigour must necessarily involve a reasonably mathematical description involving both quantum theory and special relativity. It is possible, however, to provide a qualitative account, given that we are sufficiently content to accept the notion of spin as given and to work within the framework of the old quantum theory. In particular, it is helpful to imagine that we can talk about the path of an electron in an atom in terms of Bohr–Sommerfeld orbits. These are little taught today and we shall not require to know much about them; just enough to note the idea that electron orbits corresponding to variations in what we would now call orbital quantum numbers (l values) are represented by differing degrees of ellipticity. In a totally schematic way, Figure 4.6 sketches Bohr–Sommerfeld orbits for

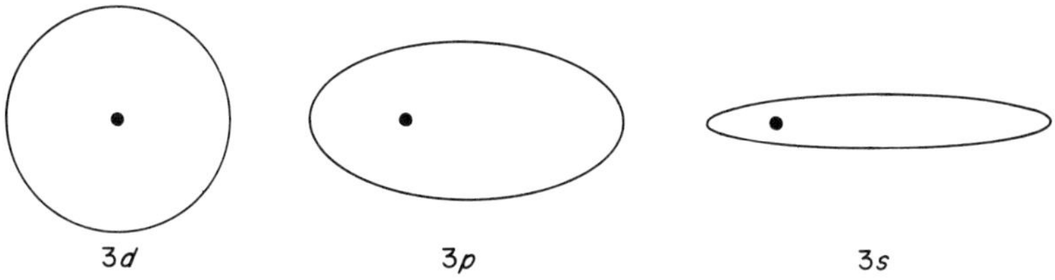

Figure 4.6 The azimuthal quantum number is represented by the degree of ellipticity in the old Bohr–Sommerfeld model.

3*s*, 3*p* and 3*d* shells. In passing, it should be emphasized that the only thing really unacceptable today about such representations is the idea that we can meaningfully talk about an electronic pathway: this objection is, of course, fundamental, being at variance with the uncertainty principle. In other respects the Bohr–Sommerfeld picture usefully summarizes other concepts with which we remain happy in the new quantum theory. In particular, the notion of electron *penetration* is well represented. As we traverse the series, 3*d* → 3*p* → 3*s*, the orbits become more eccentric and so the electron spends an increasing fraction of its time very close to the nucleus. The same idea is contained within the statistical view of current quantum theory when, as shown in Figure 2.9, we note an increasing number of subsidiary maxima in the radial waveforms of these orbitals along the same series. Indeed, a simple way of remembering the shapes of Bohr–Sommerfeld orbits is to note that for the maximum *l* value allowed for a given principal quantum number (here, 3*d*) we draw a circle and the radial distribution function shows no subsidiary maxima. As the *l* value decreases, we systematically add subsidiary inner maxima and make the Bohr–Sommerfeld orbit more eccentric. The purpose of this short review of the old quantum ideas is largely to assure the reader that the pictures we are about to study parallel a rigorous view more closely than is at first apparent. So let us now look at this view.

First recall our discussion in §4.1, and Figure 4.1. Spin-orbit coupling energies are determined by the scalar product of spin- and orbital-angular momentum operators times a radial integral like ζ. If we enquire, for example, how such energies change as we descend a group, changing from 3*d* to 4*d* to 5*d*, the product of angular momentum operators does not change at all: instead it is the *efficiency* of the process, as represented by the spin-orbit coupling coefficient. A neat trick that helps us analyse this sort of variation is to invert the picture in Figure 4.1 and consider instead the view from the electron. This is characterized by the nucleus orbiting the electron with exactly the same orbital angular momentum as the electron actually possesses with respect to the nucleus. However, the magnetic field 'seen' by the electron, and which ultimately determines the extent of coupling between orbital motion and electron spin, depends not only upon the angular velocity of the nucleus but also upon the effective charge experienced at any point by the electron. Note straightaway that this effective nuclear charge is not the same as the 'chemical' Z_{eff}, for that refers to the ionization energy of a given (but probably different) valence electron and so involves an average. Here we envisage the variation in the instantaneous charge as the nucleus completes its orbit. However, having used the reciprocal viewpoint to establish that the spin-orbit coupling coefficient depends in some way upon the nuclear charge, it is now more convenient to return to the more usual viewpoint and consider again the electron as orbiting the nucleus. If the electron is in a more penetrating orbital, or travelling a more eccentric elliptical path, it spends an increasing period of time at close quarters with the nucleus, as we have discussed. Schematically, at least, the sequence of Bohr–Sommerfeld orbits from left to right in Figure 4.6 for 3*d*, 3*p*, 3*s* electrons would suffice also for the sequence 3*d*, 4*d*, 5*d* as the reader should check. While the electron in, say, a 5*d* orbit is near the nucleus, the contribution to the spin-orbit coupling energy is greater than when it is further away: and the effect is anything but linear. Overall, the magnitude of a spin-orbit coupling parameter is dominated by that interval of time spent by the electron close to the nucleus. Spin-orbit coupling is often called an 'inner effect', because of this. The first-order relativistic theory of Dirac applied to the hydrogen atom with a Coulombic central field potential of the form,

$$V = -Ze/4\pi\epsilon_0 r \tag{4.20}$$

yields[7] the expression

$$\zeta_{nl} = \frac{Z^4 e^2 \hbar^2}{8\pi\epsilon_0 m_0^2 c^2 a^3 n^3 l(l+\frac{1}{2})(l+1)} \tag{4.21}$$

for the one-electron, spin-orbit coupling coefficient. The dependence of ζ upon the nuclear charge Ze and on the electron-nuclear distance a is evident. The appearance of this expression has led some to describe the spin-orbit coupling effect as an 'interaction with the nucleus' but that is not really a helpful view.

The extent of the dependence of the ζ coefficient on position in the periodic table is shown by the selected values given in Table 4.1: it is obviously very great indeed. Now at the same time that the spin-orbit perturbation assumes significance as we descend the table, the electrostatic Coulombic effect decreases. This follows because valence atomic orbitals become more diffuse with increasing atom size and decreasing effective nuclear charge so that, on average, any two electrons are in less close proximity. Somewhere in the periodic table, therefore, it seems possible that the magnetic

Table 4.1 Values (cm^{-1}) of spin-orbit coupling coefficients for selected atoms and ions

(a) *First period neutral atoms*

Atom	B	C	N	O	F
Valence configuration	$2p^3$	$2p^4$	$2p^5$	$2p^6$	$2p^7$
ζ_{2p} (approx)	15	30	50	70	140

(b) *First transition period ions*

Ion	Ti^{3+}	V^{3+}	Cr^{3+}	Mn^{3+}	Fe^{2+}	Co^{2+}	Ni^{2+}	Cu^{2+}
Valence configuration	$3d^1$	$3d^2$	$3d^3$	$3d^4$	$3d^6$	$3d^7$	$3d^8$	$3d^9$
ζ_{3d}	155	210	270	350	410	530	650	830

(c) *Second transition period ions*

Ion	Zr^{3+}	Nb^{3+}	Mo^{3+}	—	Ru^{2+}	Rh^{2+}	Pd^{2+}	Ag^{2+}
Valence configuration	$4d^1$	$4d^2$	$4d^3$	—	$4d^6$	$4d^7$	$4d^8$	$4d^9$
ζ_{4d}	500	670	820	—	990	1235	1615	1845

(d) *Third transition period ions*

Ion	Hf^+	Ta^+	W^+
Valence configuration	$5d^3$	$5d^4$	$5d^5$
ζ_{5d}	1335	1775	2560

(e) *Lanthanide* (+3) *ions*

Ion	Ce	Pr	Nd	Pm	Sm	Eu	Gd	Tb	Dy	Ho	Er	Tm	Yb
Valence configuration	$4f^1$	$4f^2$	$4f^3$	$4f^4$	$4f^5$	$4f^6$	$4f^7$	$4f^8$	$4f^9$	$4f^{10}$	$4f^{11}$	$4f^{12}$	$4f^{13}$
ζ_{4f}	640	750	900	—	1180	1360	—	1620	1820	2080	2470	2750	2950

coupling might overtake and even dominate the electrostatic, interelectron repulsions and so point to the need for the alternative **j–j** coupling scheme. In fact, the Coulomb terms never do acquire so small a role that this coupling scheme becomes as 'good' as the Russell–Saunders scheme is for the lighter elements; and this is one reason for the brevity of the exposition now provided.

4.8 *The j–j coupling scheme*

In this sequence we consider the magnetic coupling to outweigh the electrostatic. We briefly study the scheme by reference to the d^2 configuration once more.

For each electron, $l = 2$ and $s = \frac{1}{2}$ so, from the vector triangle rule (4.7) $j = \frac{5}{2}$ or $\frac{3}{2}$. Figure 4.7 summarizes the situation thus reached after the magnetic coupling process.

The groups of functions on the right are labelled as $[j_1, j_2]$ and, for d^2, by Hund's third rule, will have the relative energies shown. At this stage, the electrons are uncoupled so that the degeneracies of these bracketed pairs is just $(2j_1 + 1)(2j_2 + 1)$. There are, however, two ways of achieving the $[\frac{3}{2}, \frac{5}{2}]$ pairing according to the non-quantum labelling of electrons as 'first' and 'second'. A quick check shows that the total degeneracy spanned by these pairs is 100. As in Chapter 3, we must now remove those electronic

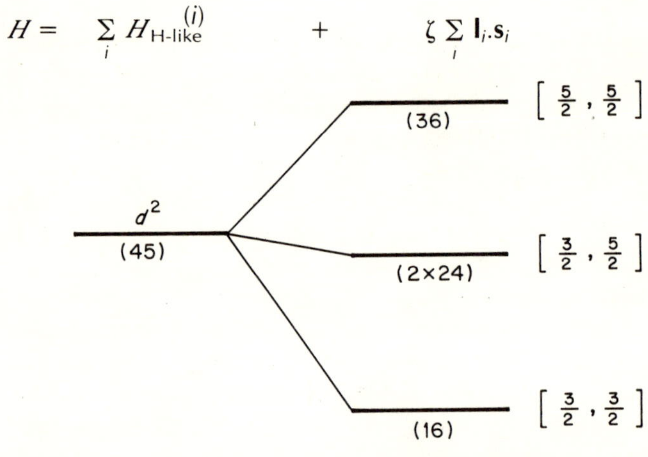

Figure 4.7 j–j coupling: the free-ion d^2 configuration split by spin-orbit coupling.

arrangements that are disallowed either by the exclusion principle or by the indistinguishability of electrons. We proceed by considering each $[j_1, j_2]$ pair in turn.

For the [3/2, 3/2] couplet, we use the vector triangle rule (4.10) to produce levels with possible J values of 3, 2, 1 and 0. We establish which ones actually arise in the case of equivalent d electrons by constructing Table 4.2, akin to Table 3.3.

Table 4.2

M_J	3	2	1		0	
$(m_j(1), m_j(2))$	~~(3/2, 3/2)~~	(3/2, 1/2)	(3/2, −1/2)	~~(1/2, 1/2)~~	(3/2, −3/2)	(1/2, −1/2)

The table is again symmetric in M_J about $M_J = 0$ so only positive M_J values are included explicitly. The microstates now refer to m_j couplets rather than to m_l, m_s quantum numbers and, once again, Pauli-forbidden microstates are crossed out twice. Any level characterized with the total angular momentum quantum number J must be complete, of course, and have $(2J+1)$ associated M_J values. Clearly the level with $J = 3$ cannot arise in the present case. On completing that with $J = 2$, we are left with only one microstate—in the box labelled $M_J = 0$. Altogether, therefore [3/2, 3/2] gives rise to states with $J = 2$ and 0 only, for equivalent electrons.

As the j values in [3/2, 5/2] differ, no ensuing microstates will violate the Pauli principle and when we disregard the order of writing the m_j values in the microstate, we obtain just one each of all four J values furnished by the vector triangle rule; namely, $J = 4$, 3, 2 and 1.

For the couplet [5/2, 5/2], the vector triangle rule suggests J values 5, 4, 3, 2, 1 and 0. It is left as a simple exercise for the reader, using a table like Table 4.2, to confirm that for equivalent electrons, only levels with $J = 4$, 2, and 0 survive.

In summary: we have reached the situation shown in Figure 4.8. A quick check shows that the total degeneracy of the stack of levels on the right equals that of the d^2 configuration on the left. Of course the reader will have observed that the situation on the right of this figure, representing as it does, the perturbation of a d^2 configuration by both magnetic and electrostatic mechanisms, is

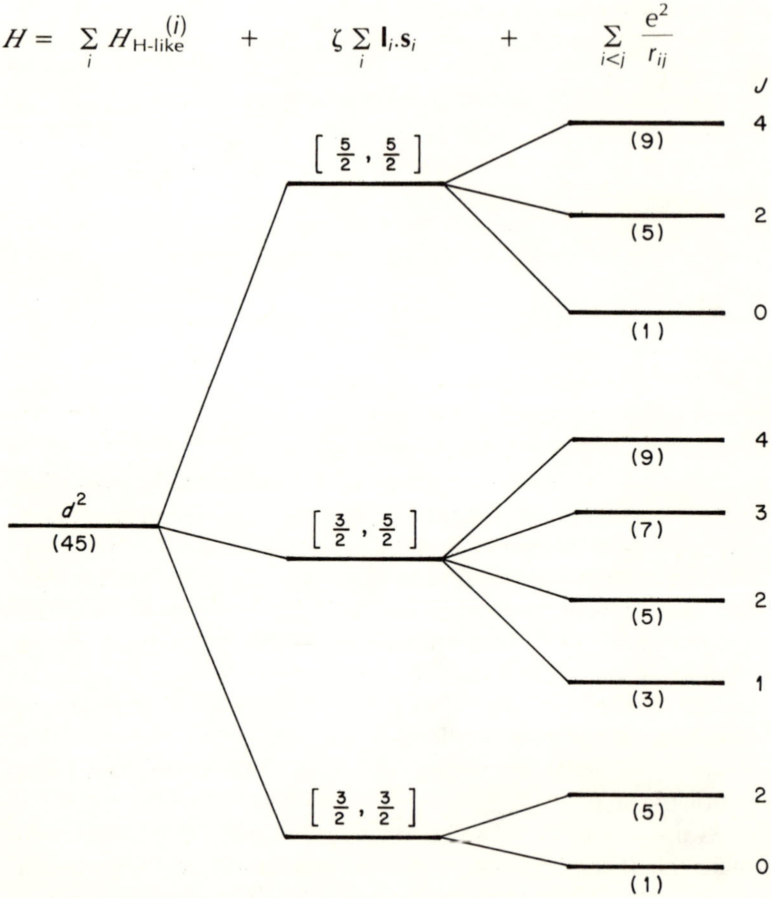

Figure 4.8 Splittting of the complete d^2 configuration in the **j–j** coupling scheme.

equivalent to that on the right of Figure 4.5. As discussed in connection with Figure 4.2, the wavefunctions resulting from either the Russell–Saunders or the **j–j** coupling scheme form eigenfunctions for J^2 and J_z but otherwise differ with respect to operators at the intermediate stages. Nevertheless for species characterized by the final total angular momenta, J, M_J, the end products of the two coupling schemes must match up.

4.9 *Intermediate Coupling*

The 'matching up' between the two coupling schemes is effected through a so-called *correlation diagram*, and for the d^2 case, this is shown in Figure 4.9. We note first that the two coupling sequences (RS and **jj**) yield the same numbers of the same J levels, as they must. The labels for these levels differ in the two schemes because different intermediate stages have provided parentage. In the level symbol $^3F_2$, for example, the nomenclature states that we have a level characterized by $J = 2$, arising from the term ^3F. The lowest-lying level with $J = 2$ on the right arose from the couplet [$\frac{3}{2}$, $\frac{3}{2}$]: and so on.

The region in the centre of the diagram is labelled 'intermediate coupling' but not to imply that a third coupling scheme is possible, for it is not. The expression is used more to indicate that deviations from the Landé interval rule are becoming energetically significant. We discuss this concept in a moment. First note that links between levels of the same J value are made according to the *non-crossing rule*. We discuss the origin of this rule in Chapter 6, being content here merely to quote and use it. It says in effect that, 'two or more states characterized in the same way with respect to some process cannot cross one another in response to a variation in the strength of that process'. A less exact but more immediate statement of the non-crossing rule is that 'states of the same symmetry cannot cross'. In both statements the word 'state' refers to any wavefunction—state, level or other. Applying the rule to the correlation of levels in Figure 4.9, we note, for example, that there are two levels with $J = 0$, three for $J = 2$ and two for $J = 4$. The correlations have been made in such a way as to obviate the need for such repeated 'links' to cross one another. (This can always be done by working, say, from the bottom of the figure, taking the first with the first, the second with the second, and so on.)

4.10 *Deviations from Landé's rule*

Representing energy levels by horizontal lines in diagrams like Figure 4.9 is commonplace and convenient. But no summarizing scheme like that tells the whole truth and can sometimes be misleading. Consider as an example, the splitting of the ^3F term by spin-orbit coupling. Instead of Figure 4.4 we might represent the spin-orbit perturbation as being gradually increased from zero

$$H = \sum_i H_{\text{H-like}}(i) + \sum_{i<j} \frac{e^2}{r_{ij}} + \lambda \mathbf{L.S} \qquad \sum_{i<j} \frac{e^2}{r_{ij}} + \zeta \sum_i \mathbf{l}_i.\mathbf{s}_i + \sum_i H_{\text{H-like}}(i) = H$$

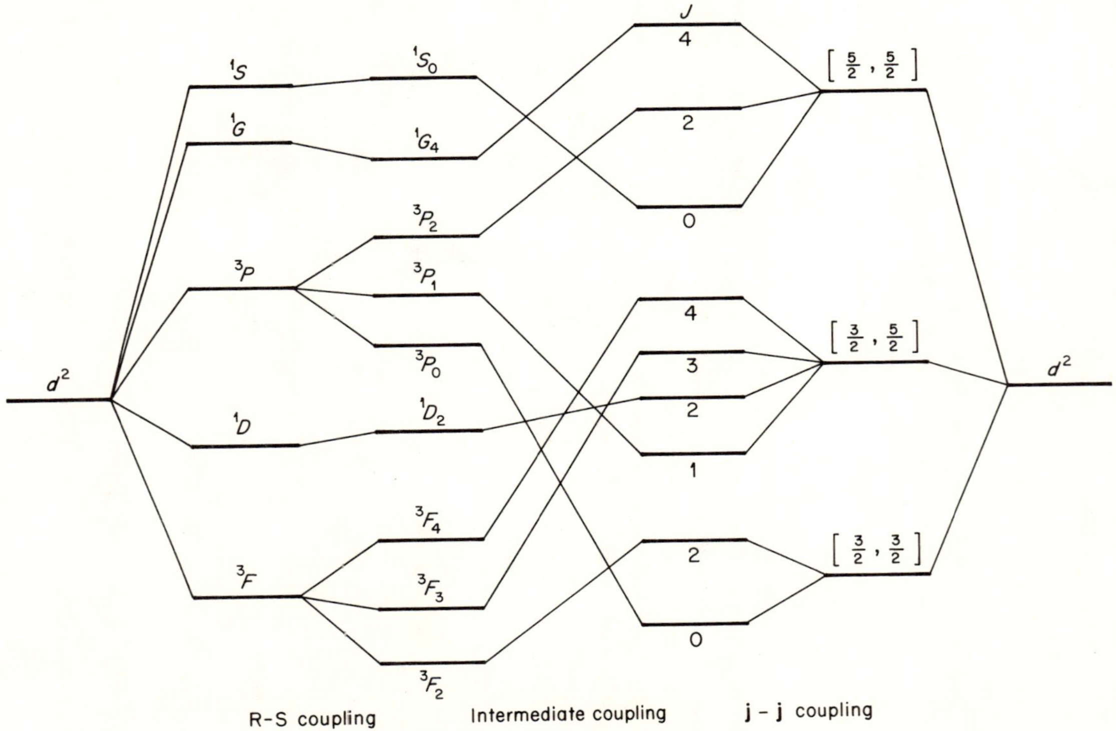

Figure 4.9 Correlation diagram showing how Russell–Saunders and **j–j** coupling schemes yield the same result eventually.

against a background of fixed Coulomb interaction, as in Figure 4.10. The energy of a level arising from any given term was calculated in (4.16): substitution of the appropriate L, S and J values into that expression yields energies of $^3F_4$, $^2F_3$ and $^3F_2$ levels as 3λ, $-\lambda$ and -4λ, respectively. In Figure 4.10, these energies are represented as the slopes of straight lines radiating from the parent term energy: at any point along the abscissa, the level energies will be given by these slopes times the current λ value. Note in passing the operation of a 'barycentre' or 'centre of gravity' rule in that summing the level energies after each is multiplied by its degeneracy yields zero. This merely serves to emphasize that the spin-orbit coupling operator was constructed to have this property—that it refers to term *splitting*.

Now with the representations in Figure 4.10 in mind, consider again a typical region of the correlation diagram in Figure 4.9. In the *limit* of vanishing spin-orbit coupling, all these levels collapse smoothly into the single entity we write as ^3F. On the other hand, in the limit of vanishing Coulombic interaction, these same levels acquire two different energies—those of the [3/2, 3/2] and the [3/2, 5/2] couplets. The relative splitting pattern of the ^3F term, as described by the Landé interval rule, therefore cannot be maintained across the correlation diagram. The splitting ratio of 4:3 on the left

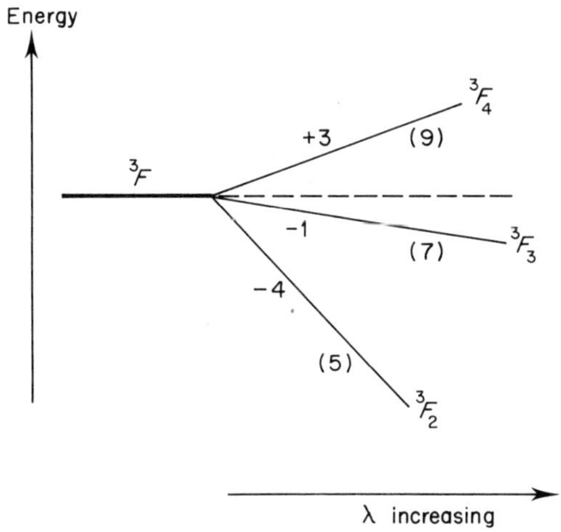

Figure 4.10 An alternative illustration of the splitting of a ^3F term by spin-orbit coupling.

becomes one of $0:5\zeta/2$. (As may be checked using (4.19) with lower-case letters.) This means that the Landé interval rule can only apply *strictly* in the limit of vanishing λ and that the slopes of the level energies in Figure 4.10 are really asymptotes and the true level energies must curve with respect to linear variations in λ. In fact, such energy-level curvatures are associated only with those states with siblings of the same J value. In the jargon of this subject, we refer to levels of the same J value having 'mixed' with one another. The concept is closely associated with the non-crossing rule we used to construct the correlation diagram. We do not pursue the matter here, however, though some clarification will be attempted in Chapter 6.

Examples of the breakdown of Landé's interval rule are found regularly throughout the level energies of the lanthanide metals. A great many terms and levels arise from f^n configurations and their study has provided much interest and complexity, although no new principles beyond those described in the last three chapters. It is common to observe levels arising from a given term in these systems to be roughly evenly split—as if they all 'repel' each other—showing that the Russell–Saunders coupling scheme is beginning to fail quantitatively. Atomic spectra falling even further into the intermediate coupling region occur for elements of the third-row transition block when λ values are very large indeed (see Table 4.1). However, nowhere in the periodic table of natural elements does the **j–j** coupling scheme offer a marked advantage over the Russell–Saunders. Deviations from either (by which is meant these energy level curvatures) are large and if quantitative results are required, a full computation of the whole set of level interactions becomes necessary. Thus Russell–Saunders and **j–j** coupling schemes are qualitatively correct, of course, but quantitatively are just limiting cases. It is simply an empirical fact that the Russell–Saunders scheme provides a reasonably accurate basis for discussion throughout so much of the periodic table.

At the end of §4.6, you were asked to discuss the spin-orbit splitting of a ^4P term arising from the configurations d^5. The levels arising are $^4P_{5/2}$, $^4P_{3/2}$, $^4P_{1/2}$ from the vector triangle rule. In first order, however, these levels are not split energetically, as we know from the hole formalism which puts d^5 as its own hole-equivalent. Thus if we consider d^5 as less-than-half full $^4P_{1/2}$ lies lowest in energy; but if as more than half full, $^4P_{5/2}$: both statements can be true simultaneously only if the splitting is identically zero—a sort of schizophrenia at the atomic level! In fact, the levels split very widely indeed and these energy shifts arise solely from higher-order processes—from the mixing of other $J = 5/2$, $3/2$ and/or $1/2$ levels into those from ^4P. A similar example of energy level splittings in d^5 arising only from higher-order processes occurs with another common one-electron operator—namely, with the ligand-field potential: the Tanabe–Sugano diagram for d^5 looks similar to those for other d^2 configurations but only arises in this way.

4.11 *The effects of applied magnetic and electric fields*

In real atoms the effects of both Coulombic and magnetic coupling are endemic. We can study the situation further, however, in a controlled way, by subjecting atoms and molecules to laboratory-determined electric and magnetic fields. The levels split up in different ways, depending upon the kind of field applied, the phenomena being referred to, respectively for electric and magnetic fields, as the Stark and Zeeman effects. We confine remarks here to a most brief review of the gross features of these phenomena.

Any level of an atom is degenerate with respect to its M_J (or m_j, as appropriate) components. It has been made clear from the beginning that terms and levels must be complete, for that merely reflects the isotropy of space. That the different M_J components belonging to any given level must have the same *energy* in the free atom is a little less obvious, perhaps.† It arises, essentially because the quantization axis (usually called z), with respect to which M_J components are characterized, is not fixed in space by any *physical* property. Therefore, since all directions are alike to an atom, no physical, observable difference can be ascribed to wavefunctions that differ only in their orientation with respect to a notional reference axis. On the other hand, if some direction can be made physically unique by experimental means, differentiation between the various M_J components becomes possible. Externally applied electric and magnetic fields provide just such a physical reference. It is conventional and convenient, though not theoretically essential, to define this axis as z and to quantize all earlier angular momenta with respect to it.

† Degenerate means having the same eigenvalue of a given operator—here J^2—not necessarily the same E value.

The energies of wavefunctions with common J value but different M_J now become different: all are distinct in a magnetic field; *not all* in an electric one—we return to this comparison in a moment. We can understand these energy differences by focusing on the magnetic moments associated with the different M_J values. Their fields are oriented parallel to, or antiparallel to, or at intervals between, the applied field in the semi-classical model and their complex precessional movements about the field differ accordingly. It is not important for us to analyse the details of those precessions for they are an artefact of the old quantum theory anyway. It is sufficient to recognize that all $(2J + 1)$ components of a level characterized by J will interact differently with an applied magnetic field and we can summarize the result by Figure 4.11. This is the *Zeeman effect*. The splitting is linearly proportional to the magnitude of the applied magnetic field, B, and the components of the state split equally by an amount labelled $g\beta_0 B$. The Bohr magneton, β_0 is just a collection of fundamental constants:

$$\beta_0 = \frac{e\hbar}{mc^2} \tag{4.22}$$

where e and m are the charge and mass of the electron, and c is the velocity of light. The remaining part of the proportionally constant, g, is called the *Landé splitting factor* or just the g *value*. For free ions—and only for free ions or atoms—the Landé g value is given by

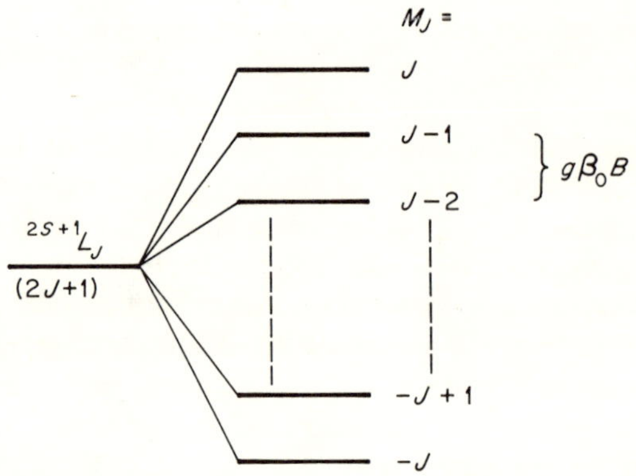

Figure 4.11 The degeneracy of the level is removed completely in an external magnetic field.

$$g = 1 + \frac{J(J+1) - L(L+1) + S(S+1)}{2J(J+1)}, \tag{4.23}$$

an expression we shall not derive here. In the experiment known as *electron spin resonance (e.s.r.)* or *electron paramagnetic resonance (e.p.r.)*, g values in atoms and molecules can be measured by first creating the level splitting by placing the sample in a magnetic field and then performing spectroscopy by irradiating the molecules with light of the appropriate frequency. As $g\beta_0 B$ for typical atoms and molecules in normal laboratory magnetic field strengths of, say, 1 tesla, lies in the range 0.1 to 2 cm^{-1}, such radiation falls in the microwave region. Absorption occurs when the microwave frequency exactly matches the energy separation between adjacent M_J components (adjacent, because the absorption follows a selection rule that transition only occurs for shifts of M_J by unity):

$$\nu = g\beta_0 B. \tag{4.24}$$

Needless to say, a satisfactory description of the e.s.r. experiment is far more demanding than this brief note and falls outside our present scope.

In reviewing the Zeeman effect, we can at least comment upon the name and symbol used for the quantum number referring to the J_z (or L_z or S_z) operator. We call M_J (or M_L or M_S) *magnetic quantum numbers* because they require an *external* magnetic field to resolve their associated energies completely. Note, in passing, that there is not just *the* magnetic quantum number M or m but the range of orbital, spin or total magnetic quantum numbers according to the context.

In an electric field, the degeneracy of a state is only partially raised. In fact the so-called *Stark effect* distinguishes between $|M_J|$ values but not their signs. For example, in an externally applied electric field, the $^3F_2$ level splits as in Figure 4.12. We do not pursue the reasons for this difference here except to note, for interest's sake, a difference between the character of magnetic and electric fields. They can both be described as vector quantities but while the electric vector is a polar vector, the magnetic one is an axial one. The formal distinction between these quantities is described elsewhere[7] but we can readily discern one immediate difference between them. Only one convention needs to be adopted to define the sense of a polar vector: the electric field vector points towards,

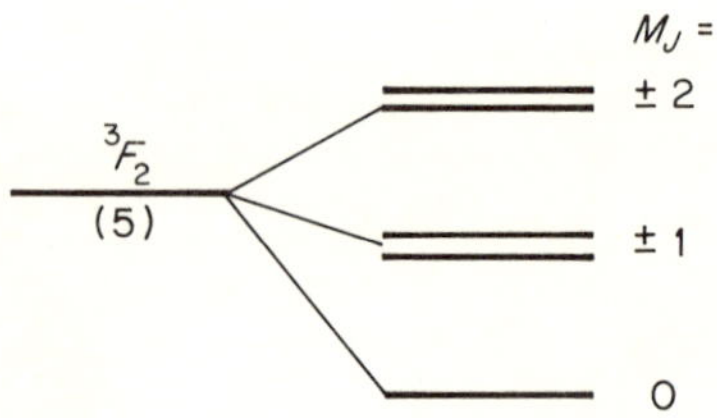

Figure 4.12 An external electric field cannot resolve the signs of M_J values.

say, a positive point charge. The axial vector requires a second, independent convention: a magnetic north pole is produced when a negative electric current rotates around a loop clockwise, say. The nature of electric and magnetic fields are intimately linked with their interaction with matter. It is a fascinating subject but we must leave it here.

4.12 *Atomic orbitals, configurations, terms, levels and states*

Here we summarize the *names* given to the various entities we have studied in connection with atomic spectroscopy.

An *orbital* is any one-electron wavefunction. In a less rigorous context we might associate a pair of electrons with an orbital but really we should distinguish these situations by use of *spin-orbitals*.

A *configuration* describes a group of many-electron wavefunctions that would arise in the absence of the Coulomb operator. For n electrons housed in an l shell, the degeneracy of a free-ion configuration is given by the combinatorial formula $^{2l}C_n$.

A free-ion *term* is a set of many-electron wavefunctions sharing common energy, and spin and orbital angular momentum properties. The term ^{2S+1}L is $(2S+1)(2L+1)$-fold degenerate. Each component can be called a *term wavefunction* or *state*.

A *level*, in the present specialist context, is a set of many-electron wavefunctions sharing common energy (and spin and orbital) and total angular momentum properties. The level $^{2S+1}L_J$ is $(2J+1)$-fold degenerate and its components are called *states*.

States are one-electron or many-electron wavefunctions that are completely specified—in the present context, with respect to spin, orbital, total and z-component of total angular momenta. Therefore, a spin-orbital is a special example of a state. Also, since the adjectival clause, 'many-electron', can include the case of 'one-electron', common sense must be exercised in the use and interpretation of these various names. As ever, it is the concepts that are important and we have to be relaxed about frequent abuses of the names.

CHAPTER 5

Antisymmetry

We leave the subject of atomic spectroscopy now but we shall return to that of angular momentum again. Here we explore a different but most fundamental aspect of quantum theory.

5.1 *The exclusion principle*

Most statements in quantum mechanics seem strange at first, being frequently counter-intuitive. But, as discussed in Chapter 1, we should not expect common sense necessarily to be helpful in matters inaccessible to common experience. After a while one becomes inured to the seemingly peculiar postulates of quantum mechanics. Beware, however, the innocent-looking assertions. It is held, for example, that particles of the same kind (electrons or protons, for example) are indistinguishable from one another. Indeed, we have made use of this rule already in following through vector coupling theory. The seemingly innocent and obviously self-evident nature of this principle of indistinguishability has enormously important repercussions. Consider first the same statement but applied to macroscopic objects. For example, and leaving aside trivial irregularities, we might say the bricks in a wall are indistinguishable, having been factory produced like peas in a pod. But that would not be correct, for we *can* distinguish the bricks if only by virtue of their different locations. In the classical world we can *always* distinguish otherwise identical objects by virtue of position and so in this sense, the classical world does not recognize the idea of indistinguishability at all. If the reader finds this point argumentative, let him understand that the real distinction being made here is between objects as they 'are' and objects as they 'appear'. Before the advent of quantum theory and all its associated philosophy, the notion of 'objects as they are' in science was totally comprehensible: indeed it squares with common sense. Today, however, we recognize that science is about the relationships between *observations* and the concept of what *is* is scientifically without definition. So we must say that the bricks in the wall can always be distinguished rather than that they are identical.

When we claim that fundamental particles are *indistinguishable*, however, we mean exactly that: not that they are identical, for we cannot pursue the notion experimentally, but that they cannot be told apart by experiment. Let us see where this fundamental postulate of quantum theory leads us.

We write ϕ here to represent a full description of the wavefunction of some fundamental particles. For example, if we refer to electrons, ϕ means a spin-orbital rather than just a description of the space part. Consider a description of some system comprising two such particles. Let their complete wavefunctions be written ϕ_1 and ϕ_2. Initially we might consider writing the simple product $\phi_1(1)\phi_2(2)$ to describe the two-particle system. A pointer to why we take a *product* could be the analogy of Lissajous figures formed by the product of two orthogonal (independent) simple harmonic oscillations: the contribution of each component in the product is independently maintained. The notion used in our simple product is such that the numbers in brackets label the particles, so we read $\phi_1(1)\phi_2(2)$ as: particle 1 is described by wavefunction ϕ_1 and particle 2 by function ϕ_2. However, we could just as well associate particle 2 with ϕ_1 and 1 with ϕ_2, as in $\phi_1(2)\phi_2(1)$. This description—which is certainly different on paper—refers to a different particle arrangement which is, however, experimentally indistinguishable from the first. The two arrangements are equally acceptable descriptions of the two-particle system and so a proper description should mention them both with equal weight. Thus we write:

$$\psi = \frac{1}{\sqrt{2}}\{\phi_1(1)\phi_2(2) \pm \phi_1(2)\phi_2(1)\}, \tag{5.1}$$

where we include the factor $1/\sqrt{2}$ just to keep the two-particle wavefunction ψ normalized (we presume ϕ_1 and ϕ_2 have already been normalized). The weightings of $\phi_1(1)\phi_2(2)$ and $\phi_1(2)\phi_2(1)$ are obviously numerically equal but they may still differ, depending upon the choice of sign. Why do we have the freedom to choose either sign? Well, the choice arises because ψ, like any wavefunction, is not itself observable. The square (or in general $\psi^*\psi$), however, corresponds to particle density—again by fundamental assertion of quantum theory. The total particle density ρ associated with either of the ψ in (5.1) is given then by integration:

$$\rho \rightarrow \int\int \phi_1^2(1)\phi_2^2(2)\, d\tau_1 d\tau_2 + \int\int \phi_1^2(2)\phi_2^2(1) d\tau_1 d\tau_2, \quad (5.2)$$

where τ_1 refers to the set of coordinates of electron 1 and τ_2 to those of electron 2. Notice that the cross terms vanish,

$$\int\int \phi_1(1)\phi_2(1) d\tau_1 \phi_1(2)\phi_2(2) d\tau_2 = 0 \quad (5.3)$$

since ϕ_1 and ϕ_2 are orthogonal functions. Altogether then, the observable particle density ρ is the same whichever sign is taken in (5.1).

The wavefunction ψ is different, nevertheless, for the different sign choice. So which is correct? The answer comes in the form of a statement about the fundamental nature of particles. It is not provable, any more than the statement that 'an orange tastes of orange' is provable. It is a question of definition. Quantum theory provides a successful account of things if we assert that particles fall into two categories: those which take the positive sign and those which take the negative. Thus:

Bosons are particles with integral spin angular momentum quantum numbers. They are associated with the positive sign in (5.1); are symmetrical with respect to particle interchange and they obey Bose–Einstein statistics. Example: α-particles.
Fermions are particles with half-integral spin. They are associated with the negative sign in (5.1); are antisymmetric with respect to particle interchange and they obey Fermi–Dirac statistics. Examples: electrons, protons.

We shall have little to say here about bosons. Almost all that we might usefully theorize about in chemistry is to do with electrons and these are fermions. Let us begin with this word 'antisymmetric'.

Fermions are associated with the minus sign in (5.1). So a two-fermion—but let us be specific and talk now about a two-electron—system is written,

$$\psi_{12} = \frac{1}{\sqrt{2}} \{\phi_1(1)\phi_2(2) - \phi_1(2)\phi_2(1)\}. \tag{5.4}$$

On interchange of the electrons we write

$$\psi_{21} = \frac{1}{\sqrt{2}} \{\phi_1(2)\phi_2(1) - \phi_1(1)\phi_2(2)\} \tag{5.5}$$

and it is clear that

$$\psi_{21} = -\psi_{12}. \tag{5.6}$$

That is what 'antisymmetric' means; that the many-fermion wavefunction changes sign on interchange of any two particles. Obviously bosons are symmetric with respect to particle interchange, for the minus sign in (5.6) originates from that in (5.4).

Suppose now that we wished to construct a similar many-electron wavefunction for a system with three electrons. Following the same line of argument we would need all permutations of the three electrons within three spin-orbitals ϕ_1, ϕ_2, ϕ_3; and with each such permutation (e.g. $\phi_1(2)\phi_2(3)\phi_3(1)$) we must associate a sign, such that the complete, many-electron wavefunction ψ_{123} changes sign on interchange of *any* two electrons. Put this way, the reader will see that the task of construction is nontrivial. Had we posed the problem of a six-electron wavefunction we would find it quite hard even to write down all the permutations, let alone determine their signs. There is a method of presenting antisymmetrized wavefunctions, however, which systematizes these processes, if explicitly necessary, but generally obviates the need to tackle the problem anyway. And there is another bonus too.

Look at the *pattern* of numbers appearing in the bracket of (5.4); (11 × 22) − (12 × 21), and ask where you have seen it before. From determinants, surely. Slater proposed the use of *determinantal wavefunctions* as a means of correctly antisymmetrizing product functions. The two-electron determinant for ψ_{12} is thus written.

$$\psi_{12} = \frac{1}{\sqrt{2}} \begin{vmatrix} \phi_1(1) & \phi_1(2) \\ \phi_2(1) & \phi_2(2) \end{vmatrix} \tag{5.7}$$

and the reader will check that this is identical to that in (5.4). There appears to be no agreed convention with respect to transposition of the rows and columns in these determinants: nothing physical depends on which way they are written provided, of course that one is consistent.

There is not too much to be gained writing (5.7) rather than (5.4). The advantages become manifest as the number of electrons increases, for we write the general *n*-electron determinantal function as

$$\psi_{1\ldots n} = \frac{1}{\sqrt{n!}} \begin{vmatrix} \phi_1(1) & \phi_1(2) & \ldots & \phi_1(n) \\ \phi_2(1) & \ldots & & \cdot \\ \cdot & & & \cdot \\ \cdot & & & \cdot \\ \cdot & & & \\ \phi_n(1) & \ldots & & \phi_n(n) \end{vmatrix} \tag{5.8}$$

with the same facility as we wrote (5.7). However, the mathematical properties of determinants now ensure the correct antisymmetrization of the total wavefunction. First, recalling the manner in which a determinant is expanded—element times cofactor, etc.—we find that *all* permutations of the electrons amongst the spin-orbitals are implied within the determinant. Secondly, if *any* two electrons are interchanged, corresponding here to the interchange of the corresponding columns in the determinant, the sign of the determinant is reversed. So this clever presentation automatically ensures—though usually implicitly—the correct antisymmetry of the complete many-electron wavefunction. Even the normalization constant is established by the rules of determinantal expansion.

The properties of determinants tell us one more thing of great interest. If any two rows (or columns, of course) of a determinant are identical, the determinant vanishes. That equality could arise if any two of the spin-orbitals ϕ were identical. Putting the two ideas together, we conclude that if any two electrons in a many-electron wavefunction are associated with the same spin-orbital—that is, if any two electrons are characterized in the same way—the amplitude of that many-electron function vanishes identically. This is, of course, *Pauli's Exclusion Principle*. The same result does not emerge for bosons for, without the minus sign in (5.4), the equivalence with the determinantal form (5.7) would not exist. Given the principle of electron indistinguishability, the exclusion principle and the antisymmetry requirement are one and the same thing. In passing, we can see why fermions and bosons behave according to different statistical patterns: in the loose language of ordinary chemistry, we may place a maximum of two electrons in any one orbital while there is no such restriction on the number of bosons to be placed in any one-boson wavefunction.

In summary: an extremely convenient, though non-mandatory, way of ensuring that a many-electron wavefunction is properly antisymmetrized is to construct the Slater determinantal wavefunction. Typographically, (5.8) may be replaced by the abbreviated form,

$$\psi_{1\ldots n} = |\phi_1\phi_2\ldots\phi_n|, \tag{5.9}$$

and the presentation of spin-orbitals within vertical bars then signifies the determinantal form complete with normalization factor.

5.2 *Antisymmetrized triplets and singlets*

When striking out microstates in Table 3.3, it was repeatedly emphasized that we were merely engaged upon a 'counting game'. In particular, no heed was paid to *which* microstate was struck out, from a given location in the table. For example, there are entries corresponding to $M_L=3$, $M_S=0$—$(\overset{+}{2}\ \overset{-}{1})$ and $(\overset{-}{2}\ \overset{+}{1})$—and one was accounted for by the ^1G term, the other by ^3F. But which? Well that is actually a somewhat difficult problem to solve starting from where we are in this book and, in any case, is not especially important for an introductory understanding of our topic. Instead we can study a related problem which actually has great generality, as will become apparent in due course.

Consider a system—atom *or* molecule—with two electrons distributed between two 'chemical' orbitals: for example; the states of the helium atom derived from the configuration $1s^1\ 2s^1$. In what follows, spin-orbit coupling is explicitly neglected.

Each 'chemical' orbital is twofold degenerate—ϕ_α, ϕ_β—and so our problem concerns a distribution of two electrons amongst four spin-orbitals: the total degeneracy is then given by $^4C_2 = 6$. To maintain the generality of the exercise as far as possible, we label the orbital (spatial) parts of the two types of function as *a* and *b* and, in Figure 5.1, identify those six distinct electronic arrangements.

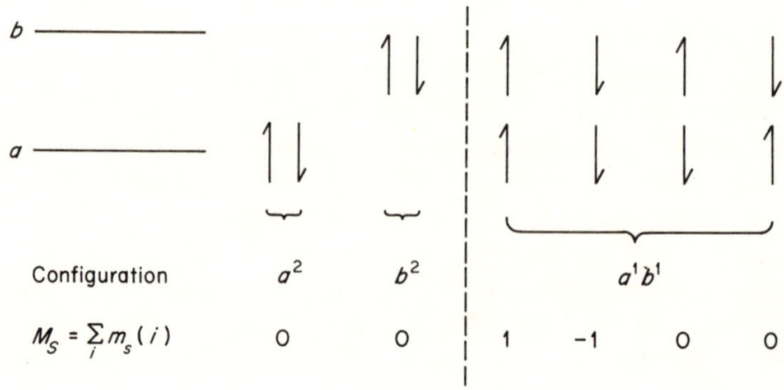

Figure 5.1 Two electrons distributed between two (chemical) orbitals.

The two configurations on the left obviously describe spin-singlets and are generally of little interest to us here. The four remaining arrangements, all belonging to the configuration a^1b^1, provide material for the study of a seminal issue. The generality of the situation derives from its description as being concerned with one electron in *each* of two unspecified (chemical) orbitals.

First we enquire what spin label will attach to the states arising from this configuration. Characterizing the total M_S values under conditions of vanishing coupling, as in the figure, we observe immediately that the $\alpha\alpha$ and $\beta\beta$ arrangements obviously belong to a spin triplet (S=1) term. The spin degeneracy of a spin triplet ($2S$+1) is 3, of course, for that is why it is so called: M_S must take the values 1, 0, −1, as usual. So one of the electronic arrangements $a\alpha b\beta$, or $a\beta b\alpha$ with M_S=0 must be struck out in completing the triplet components. That remaining then stands alone and so implies a spin singlet (S=0). So we know that one spin-singlet and one spin-triplet arise from the configuration a^1b^1: total degeneracy 3+1=4. Now we ask the inevitable question: which of the two right-hand arrangements in Figure 5.1 belongs in the triplet and which is the singlet?

The answer can be found rather quickly if we take advantage of the neglect of spin-orbit coupling (and if otherwise, the singlet and triplet labels would be less useful anyway) to factorize the wavefunctions. The independence of space and spin coordinates allows us to write the two-electron wavefunctions we seek as a simple product of space and spin parts. Consider that function describing the $\alpha\alpha$ microstate in Figure 5.1. For the space part, we ignore spin and argue that placing one electron in a and one in b can be done in two ways. So we form the combination,

$$\text{space part} \rightarrow \frac{1}{\sqrt{2}} \{a(1)b(2) \pm a(2)b(1)\}, \tag{5.10}$$

where as usual the surd normalizes the final function.

We cannot yet decide upon the sign in this function, for the antisymmetry principle refers to the total wavefunction and not just the space part. Now we construct the corresponding spin function by ignoring the spatial part. One electron is in spin state α but then so is the other, so the spin function is completely defined by the single product

$$\text{spin part} \rightarrow \alpha(1)\alpha(2), \tag{5.11}$$

for the product obtained on electron interchange is identical in every respect other than by marks on paper, which are irrelevant. Also, as interchange of electrons leaves this function unchanged, we know the spin function, at least, is symmetric. The complete function we seek is obtained as the product of space and spin parts and so the space part must be antisymmetric in order that the whole function be antisymmetric with respect to electron interchange:

$$\text{complete function} \rightarrow \frac{1}{\sqrt{2}} \{a(1)b(2)-a(2)b(1)\}\, \alpha(1)\alpha(2). \tag{5.12}$$

Let us introduce the nomenclature ψ_T^1, ψ_T^0, ψ_T^{-1} and ψ_S^0 to represent the total, two-electron, antisymmetrized functions belonging to the spin-triplet with $M_S = 1, 0\ -1$ and the spin-singlet, with $M_S = 0$, respectively. We then know that these functions must take the following forms

$$\psi_T{}^1 = \frac{1}{\sqrt{2}} \{a(1)b(2) - a(2)b(1)\}\alpha(1)\alpha(2) \tag{5.13}$$

$$\psi_T{}^0 = \frac{1}{\sqrt{2}}\{a(1)b(2) - a(2)b(1)\} \frac{1}{\sqrt{2}} \{\alpha(1)\beta(2) + \alpha(2)\beta(1)\} \tag{5.14}$$

$$\psi_T{}^{-1} = \frac{1}{\sqrt{2}}\{a(1)b(2) - a(2)b(1)\}\beta(1)\beta(2) \tag{5.15}$$

$$\psi_S{}^0 = \frac{1}{\sqrt{2}}\{a(1)b(2) + a(2)b(1)\}\frac{1}{\sqrt{2}}\{\alpha(1)\beta(2) - \alpha(2)\beta(1)\} \quad (5.16)$$

The reasoning goes as follows. We have just established (5.13) in detail and the form of (5.15) follows immediately by substituting β for α in the discussion. Next we recognize that a spin-triplet means a *spin*-triplet. That is, it is a triplet by virtue of its spin properties: there are three different arrangements possible for the electron spins. As the spin and space parts are independent and so form as factors of the whole, the space part must be *common* to all three spin parts. Therefore the sign appearing in the space parts of all three components of the spin triplet are the same. Now the spin function of ψ_T^0 involves one electron as α and one as β, and so both combinations must be included; hence the linear combination shown in (5.14). The positive sign occurring in this spin combination is forced by the negative sign just established in the space part because one or another, but not both, space and spin parts must be antisymmetric if the whole is to be antisymmetric. Finally, the function shown for the spin-singlet in (5.16) is established by similar reasoning to that just employed for ψ_T^0 together with a recognition that ψ_S^0 and ψ_T^0 must be different.

In conclusion, therefore, it is usual to call the spin triplet, antisymmetric; and the spin singlet, symmetric: these adjectives are understood to refer to the *space* parts of these functions, of course. In reply to our question about which of the microstates with M_S=0, in Figure 5.1, belongs to the triplet and which the singlet, we observe that both belong to both! This follows essentially because the microstates $(\overset{+}{a}\ \overset{-}{b})$, and $(\overset{-}{a}\ \overset{+}{b})$ are not individually antisymmetric with respect to electron interchange: thus $(\overset{+}{a}(1)\ \overset{-}{b}(2) \rightarrow (\overset{+}{a}(2)\ \overset{-}{b}(1))$ which is not the same, even barring a sign, as $(a(1)\ b(2))$. The combination of the microstates implied in (5.14) and (5.16) ensures the antisymmetry we require. We can rewrite these functions in an unfactorized fashion by juxtaposing those parts with common arguments (i.e. common electron numbers):

$$\psi_T^1 = \frac{1}{\sqrt{2}}\{\overset{+}{a}(1)\overset{+}{b}(2) - \overset{+}{a}(2)\overset{+}{b}(1)\} \quad (5.17)$$

$$\psi_T^0 = \frac{1}{2}\{[\overset{+}{a}(1)\overset{-}{b}(2) - \overset{+}{a}(2)\overset{-}{b}(1)] + [\overset{-}{a}(1)\overset{+}{b}(2) - \overset{-}{a}(2)\overset{+}{b}(1)]\} \quad (5.18)$$

$$\psi^{-1}_T = \frac{1}{\sqrt{2}}\{\bar{a}(1)\bar{b}(2) - \bar{a}(2)\bar{b}(1)\} \qquad (5.19)$$

$$\psi^0_S = \frac{1}{2}\{[\overset{+}{a}(1)\bar{b}(2) - \overset{+}{a}(2)\bar{b}(1)] - [\bar{a}(1)\overset{+}{b}(2) - \bar{a}(2)\overset{+}{b}(1)]\}. \qquad (5.20)$$

Yet a third way to write these functions, this time in terms of Slater's determinants, is,

$$\psi^1_T = \left|\overset{+}{a}\overset{+}{b}\right| \qquad (5.21)$$

$$\psi^0_T = \frac{1}{\sqrt{2}}\{\left|\overset{+}{a}\bar{b}\right| + \left|\bar{a}\overset{+}{b}\right|\} \qquad (5.22)$$

$$\psi^{-1}_T = \left|\bar{a}\bar{b}\right| \qquad (5.23)$$

$$\psi^0_S = \frac{1}{\sqrt{2}}\{\left|\overset{+}{a}\bar{b}\right| - \left|\bar{a}\overset{+}{b}\right|\}, \qquad (5.24)$$

and *now* we can reply that the triplet involves the positive combination of the last two microstates in Figure 5.1, while the singlet involves their negative combination. This view is not as revealing, however, as that yielded by the original trio of equations; namely, that the triplet is (space) antisymmetric while the singlet is symmetric.

5.3 *Coulomb and exchange integrals*

We have established that the configuration a^1b^1 gives rise to a spin triplet and singlet and we have constructed antisymmetric product functions that form a basis for them. Now we investigate the energies of these terms. Notice that we can be more ambitious in our study of this simple configuration than with the d^n systems of preceding chapters. For those species we made no examination of the wavefunctions and comments about energy were limited to the use of Hund's rules.

For the a^1b^1 configuration, the position reached so far is summarized in Figure 5.2. Since we have deliberately kept the

$$H = \sum_{i}^{2} H(i) \quad + \quad \frac{e^2}{r_{12}}$$

One-electron parts — Coulomb electrostatic coupling

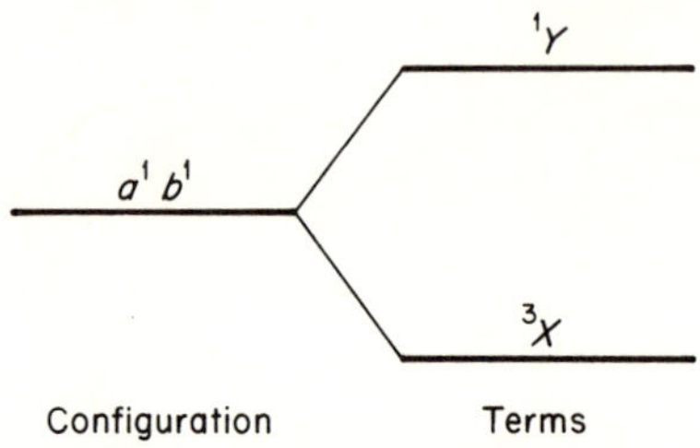

Figure 5.2 Spin-singlet and -triplet terms arising from the configuration a^1b^1.

discussion general, we use the *ad hoc* labels X and Y to represent the (different) space parts of the triplet and singlet terms. As the ^1Y and ^3X terms form eigenfunctions of the Hamiltonian operator H we know that the energy of the ^1Y term, for example, is

$$E(^1Y) = \langle ^1Y | H | ^1Y \rangle, \tag{5.25}$$

since the term wavefunctions have been made orthonormal. Breaking this down into parts, we get

$$E(^1Y) = \left\langle ^1Y \left| \sum_{i}^{2} H(i) \right| ^1Y \right\rangle + \left\langle ^1Y \left| \frac{e^2}{r_{12}} \right| ^1Y \right\rangle \tag{5.26}$$

and so the *change* in energy brought about by the Coulombic interaction is just

$$\Delta E\,(^1Y) = \left\langle ^1Y \left| \frac{e^2}{r_{12}} \right| ^1Y \right\rangle$$

$$= \left\langle \frac{1}{\sqrt{2}}\{a(1)b(2) + a(2)b(1)\}\,\text{spin} \left| \frac{e^2}{r_{12}} \right| \frac{1}{\sqrt{2}}\{a(1)b(2) + a(2)b(1)\}\text{spin} \right\rangle$$

$$= \frac{1}{2} \left\langle a(1)b(2) + a(2)b(1) \left| \frac{e^2}{r_{12}} \right| a(1)b(2) + a(2)b(1) \right\rangle \langle \text{spin} | \text{spin} \rangle. \tag{5.27}$$

The last line follows because the *Coulomb operator is a space-only operator* which thus leaves the spin function unchanged. Therefore the whole integral factorizes into a space part involving e^2/r_{12}, and a spin 'overlap integral'. Clearly, had the spin functions in bra and ket of the original integral been different, the total answer would vanish by spin orthogonality: otherwise, as here, the spin integral is unity. Continuing with the evaluation, then, we find

$$\Delta E(^1Y) = \frac{1}{2}\{[\langle a(1)b(2)\left|\frac{e^2}{r_{12}}\right| a(1)b(2)\rangle + \langle a(2)b(1)\left|\frac{e^2}{r_{12}}\right| a(2)b(1)\rangle] + [\langle a(1)b(2)\left|\frac{e^2}{r_{12}}\right| a(2)b(1)\rangle + \langle a(2)b(1)\left|\frac{e^2}{r_{12}}\right| a(1)b(2)\rangle]\}, \tag{5.28}$$

by multiplying out the space integral in (5.27). Now note that the integral pairs appearing within each square bracket are identical because the electron *numbers* are just dummy indices. If we write 2 for 1, and 1 for 2 consistently, throughout the second integral, it becomes the first integral: similarly, the fourth integral is the same as the third. Therefore,

$$\Delta E(^1Y) = \langle a(1)b(2)\left|\frac{e^2}{r_{12}}\right| a(1)b(2)\rangle + \langle a(1)b(2)\left|\frac{e^2}{r_{12}}\right| a(2)b(1)\rangle. \tag{5.29}$$

These integrals are different, however, for electrons are interchanged in, say, the ket of the second integral but not in the bra. At this point, it is more transparent to write the integrals in the more conventional form, collecting together parts referring to the same arguments (electrons):

$$\Delta E(^1Y) = \int\int \frac{e(a^*a)_1.e(b^*b)_2}{r_{12}}\,d\tau_1 d\tau_2 - \int\int \frac{e(a^*b)_1.e(b^*a)_2}{r_{12}}\,d\tau_1 d\tau_2. \tag{5.30}$$

A slight change in nomenclature has been made here: all numbers on the right appearing as suffixes label electron coordinates. In the first integral, e^2 has been divided so that $e(a^*a)_1$ may be seen as an electron density at a given point; similarly $e(b^*b)_2$. So this integral describes the quantum equivalent of a classical Coulomb repulsion between two negative charge densities. Accordingly the integral is called a *Coulomb integral.* It is given the symbol J or $J(1,2)$. Context should always prevent confusion with the quantum number for total angular momentum! Looking at the second integral, we find no simple classical analogue—it appears to have something to do with repulsion between overlap charge densities or the like. It is a purely quantum phenomenon and, as it arises from the exchange of electrons between orbitals, it is called an *exchange integral* and given the symbol K or $K(1,2)$. It occurs ultimately because of the antisymmetry principle and the indistinguishability of electrons. In summary, then, we have

$$\Delta E(^1Y) = J(1,2) + K(1,2). \qquad (5.31)$$

Now we derive a similar expression for the energy of the spin-triplet term. For this, we consider any one of the three M_S components, for they are degenerate under any Hamiltonian which does not include externally applied electric or magnetic fields. Furthermore, since the derivation will follow that given for the spin singlet, step for step, comments following (5.27) also apply and we get a non-vanishing result only if bra and ket involve the same spin function; but then it does not matter which. In fact, the calculation differs from that above in only one respect; namely, the sign in the space part of the spin-triplet wavefunction is now negative. The upshot is that

$$\Delta E(^3X) = J(1,2) - K(1,2). \qquad (5.32)$$

Let us sketch the results conveyed by (5.31) and (5.32). To do this it is useful first to recognize that the Coulomb and exchange integrals are normally positive quantities. This is illustrated by noting that a closer approach of the two charge clouds in the J integral—and for the two entities in the K integral—implies a greater repulsion and hence a more positive energy. Separation to infinity would result in those integrals tending to zero. With this as background, we construct Figure 5.3. Generally, Coulomb integrals are larger than the corresponding exchange integral and this is to be expected when we recognize that products like a^*a are between functions in the same region of space ($=a^2$ in the case that

$$H = \sum_{i}^{2} H(i) + \frac{e^2}{r_{12}}$$

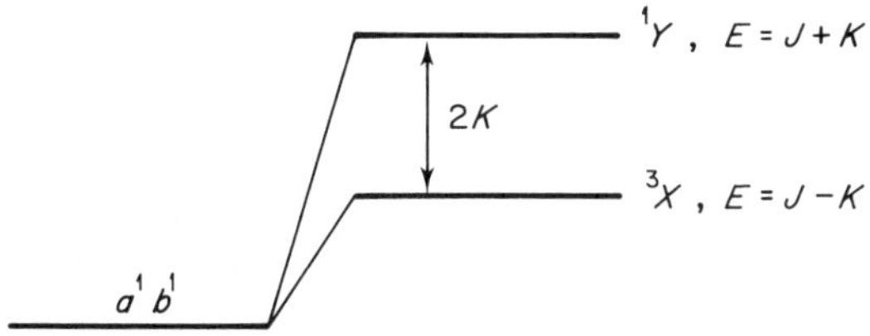

Figure 5.3 The splitting of the spin-singlet and spin-triplet terms arises from the exchange energy K.

a is real), while a^*b involves functions which maximize in different regions of space. Given this, we expect, and find, that $J-K$ is positive, so that both the triplet and the singlet are raised in energy above the parent configuration by the Coulomb interaction: this operator does not give rise to a baricentre rule. So, the proximity of the two electrons in the system leads to an increase in energy and may therefore be called a *repulsion*. If we ignore the exchange energy and set $K = 0$ for the moment, we find that both spin-triplet and singlet are raised in energy to the same extent—by J units. With respect to *that* result, the contribution of the exchange term is to split the different spin states by $2K$ units, raising the energy of one and lowering the energy of the other. In the case of the ^3X term, the lowering of the energy from the average, which originates only in the sign of (5.32) or (5.12), corresponds to the exchange *effect* for this term being *attractive*. It is as if the electrons in the spin-triplet term attract one another a bit as well as repel. This is clearly a quantum phenomenon! Of course, the exchange effect in the singlet term energy is in the opposite sense—which is why we used the word 'effect' here rather than integral. What is more important to note, however, is that the singlet and triplet terms are separated only by the exchange energy and the reader will note the emergence, in this case anyway, of Hund's first rule.

A rule of thumb that may be offered on this point is that those electronic arrangements characterized by the maximum number of parallel spins will be relatively more stable than others, by virtue of the fact that more pairs of equivalent (parallel) electrons may be interchanged amongst themselves. An example of the use of this rule in inorganic chemistry texts concerns the exchange contribution to the variation in ionization energies as a *p* shell is progressively filled. Consider just one case. The first ionization potential of the configuration p^3 involves the destruction of two pairs of parallel electrons (for the atomic ground state involves parallel electrons, one in each *p* orbital): on the other hand, the first ionization of p^4 involves the removal of the minority-spin electron because (a) the lost electron is expected to come from the more crowded or 'repulsive' environment of the doubly filled orbital and (b) no exchange energy is sacrificed thereby as the number of pairs of parallel spins remains unchanged. Since these give rise to a relative stabilization, as we have seen, the fewer parallel pairs destroyed, the better.

CHAPTER 6

Bases

6.1 *The expansion theorem*

An extremely important idea in quantum mechanics is contained within the so-called *expansion theorem* or *superposition principle*. Although this can be a very formal affair, it need not be and we shall consider it from two slightly different viewpoints—neither with much rigour—and then move on to exploit the concept. In the first, and introductory, approach, we state the theorem head on, though in a rather imprecise way.

Almost **any** function of a set of variables can be expanded in a **complete** basis of eigenfunctions, of the same variables and with the same boundary properties, of **any** operator.

By attending to those words in bold in particular, we can fill out the statement and see what it means. A well known example of the theorem—but it is to be emphasized, a fairly restricted one—is provided by the Fourier series. Consider for the moment, a function of only one variable, say $f(x)$. This can be *almost* **any** function. Strictly, it must be continuous, for example, and single-valued: a so-called function of class Q. Actually, *finite* discontinuities are permissible provided that there is not an infinite number of them within the variable range to be considered for the expansion. One such example, which also serves to illustrate the whole proposition, is the square wave shown in Figure 6.1(a).

The expansion theorem allows free choice of operator so suppose we chose $\partial^2/\partial x^2$. Eigenfunctions for this are either a set of sine functions or cosines or, indeed, their combination as an exponential:

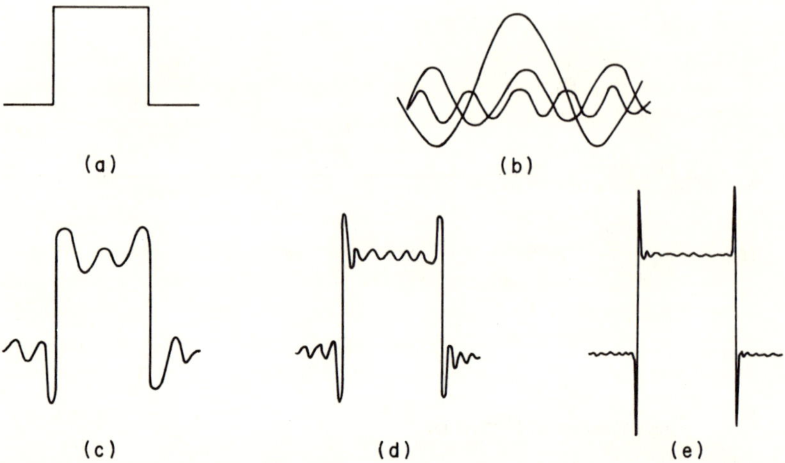

Figure 6.1 A square wave (a) may be expanded as a superposition of sine functions (b). The accuracy increases with the number of terms taken in the series expansion—(c) to (e).

$$\frac{\partial^2}{\partial x^2} \sin nx = -n^2 \sin nx \tag{6.1}$$

$$\frac{\partial^2}{\partial x^2} \cos nx = -n^2 \cos nx \tag{6.2}$$

$$\frac{\partial^2}{\partial x^2} e^{nx} = n^2 e^{nx} \tag{6.3}$$

Of course, we have chosen an operator which acts upon only one variable: it is intuitively obvious that we cannot expand a function of x in terms of functions of y, etc. More generally than the Fourier expansions above, we write

$$f = c_1\phi_1 + c_2\phi_2 + \ldots + c_n\phi_n + \ldots \tag{6.4}$$

where the set of ϕ, written as $\{\phi\}$, satisfy some eigenvalue equation:

$$\hat{O}\,\phi_i = a_i\phi_i. \tag{6.5}$$

While the operator can be selected with total freedom—though some choices are more *convenient* than others, as shall see—the associated eigenfunctions must involve the appropriate variables and boundary properties.

We refer to the eigenvectors, or wavefunctions, of the operator as a *basis*. In Figure 6.1(b) we choose to expand the square wave as a superposition of sine functions: the sine functions provide the basis of the expansion. It cannot be repeated too often, however, that nothing physical attaches to the chosen basis functions: they are selected for reasons of convenience and are virtually arbitrary.

This idea of convenience is clarified a little when we consider the word **complete** in the theorem statement. In the case of the sine functions of (6.1), for example, it is obvious that n can take any value whatsoever. However, some restriction on the possible values of n derives from the additional—usual but not essential—orthogonality requirement. The reasons for this will become clearer in our second approach to the expansion theorem in the next section. The upshot of the orthogonality—which effectively means 'independence'—of the basis functions in the case of a sine expansion, for example, is that n takes only positive, integer values. The reader may care to check that $\int_0^{2\pi} \sin x \sin 2x \, dx$, for example, vanishes within the repeat (0 to 2 π), while $\int_0^{2\pi} \sin x \sin 1.2x \, dx$ does not. And sine functions with the negative integer arguments are identical to those with positive ones. So in Figure 6.1(b) we see a superposition of only integral sine functions. There are still infinitely many of them, of course, and so the expansion (6.4) is infinitely long. That is part of the notion of completeness here. In Figure 6.1(c), (d) and (e) are sketched the results of optimal expansions—that is, those with 'best-fit' expansion coefficients—within limited or restricted bases; say, with 100, 10 000 and 50 000 terms, respectively. Incidentally, notice how the expansion takes care of the discontinuity of the square wave. By including a large contribution—large c_i in (6.4)—of the high harmonics (n large) the vertical rise can be increasingly well reproduced but only at the expense of the 'spikes' in the trace. However, these spikes have an ever-decreasing area and so the square wave will be reproduced in the limit, albeit with infinitesimally thin spikes.

The question arises at this point about how many terms are required in any given expansion. Well it all depends. . .! Suppose that somehow we have produced a perfect square wave as in Figure 6.1(a) from an electronic signal generator and then we feed the signal through a variety of 'hi-fi' audio amplifiers and thence to an oscilloscope display. We could imagine that the sort of trace shown in Figure 6.1(c) is what is obtained from a cheap piece of equipment while those in Figure 6.1(d) and (e) reveal the benefits of going up-market. Somewhere between Figure 6.1 (d) and (e) we extend the number of contributions to the sum in (6.4) beyond, say, 20 000. Since the human ear can only respond to 20 kilohertz at best, a hi-fi amplifier that does as well as Figure 6.1(e) is a waste of money—though it may 'please any passing bat' (with thanks to Flanders and Swan!). The point here is that expansions are mostly used in quantum mechanical computations to provide approximate answers and the degree of approximation must be tailored to sensible ends. The analogue of the cost of the hi-fi equipment in our example, is the effort involved in computing the expansion coefficients in (6.4). Of course, if we can choose an operator and associated basis that converges quickly towards our goal then that is a sensible and convenient way to proceed. Sometimes this involves a trade-off between convergence rate and computation time for each coefficient, as we shall see later in this chapter.

As to the question of how these expansion coefficients are calculated, a formal expression can be established very easily indeed. Thus, we multiply (6.4) throughout by ϕ_n^* and integrate:

$$\langle\phi_n|f\rangle = c_1\langle\phi_n|\phi_1\rangle + c_2\langle\phi_n|\phi_2\rangle + \ldots + c_n\langle\phi_n|\phi_n\rangle + \ldots \quad (6.6)$$

If members of the bases are orthogonal, all but $\langle\phi_n|\phi_n\rangle$ vanish on the right and since we can always work with a normalized basis, this term is just unity. Therefore,

$$c_n = \langle\phi_n|f\rangle \quad (6.7)$$

gives an expression for the (arbitrary) n^{th} coefficient in the expression. Of course, the integral $\langle\phi_n|f\rangle$ must still be evaluated and that requires time and effort.

6.2 *Hilbert space*

Now we look at essentially the same ideas but in a somewhat more formal way that simultaneously illuminates our understanding of the structure of quantum mechanics itself. First let us revise a few features of vectors in Euclidean space.

We normally think of a *vector* in ordinary space as a *directed line segment* possessing the properties of *magnitude* and *direction*:

we draw an arrow with a head and tail. It is rather more revealing to define such vectors by their involvement in certain mathematical operations because that way leads to generalization and great power. We define *scalars* in Euclidean space as the set of all *real numbers*. Then, multiplication of a vector $\mathbf{v}$ by a scalar a gives a new vector $a\mathbf{v}$ whose direction is the same as that of $\mathbf{v}$ but whose magnitude is $|a|$ times the magnitude of $\mathbf{v}$: and we note that negative scalars reverse the direction of the vector. Vector addition, $\mathbf{v}_1 + \mathbf{v}_2$, is performed by placing the tail of one vector $\mathbf{v}_2$ at the head of the other $\mathbf{v}_1$ and then constructing the directed line segment from the tail of $\mathbf{v}_1$ to the head of $\mathbf{v}_2$. Combining scalar multiplication and vector addition yields the *rule* for forming *linear combinations of vectors*:

$$\mathbf{v} = a_1\mathbf{v}_1 + a_2\mathbf{v}_2. \tag{6.8}$$

We can also construct the *inner- or dot-product*, $\mathbf{v}_1.\mathbf{v}_2$ which is given, by definition, as

$$\mathbf{v}_1.\mathbf{v}_2 = |\mathbf{v}_1|.|\mathbf{v}_2|\cos\theta_{12} \tag{6.9}$$

where θ_{12} is the angle between the two vectors. The result is a scalar and so the dot product is often referred to as the *scalar product*. In terms of a geometric construction, the inner product is obtained as the magnitude of one vector times the magnitude of the projection of the other upon it—here either $|\mathbf{v}_1|$ times $|\mathbf{v}_2|\cos\theta_{12}$ or $|\mathbf{v}_2|$ times $|\mathbf{v}_1|\cos\theta_{12}$. The scalar product of a vector with itself is called the *norm* of the vector and, apart from the trivial case of a null vector, is always a positive real number.

$$\text{norm } \mathbf{v} \equiv \mathbf{v}.\mathbf{v} = |\mathbf{v}|^2 \geqslant 0. \tag{6.10}$$

Two vectors are said to be *orthogonal* if they are perpendicular to one another. Hence, from (6.9), the scalar product of orthogonal vectors vanishes. *Orthonormal* vectors are then defined as a set of orthogonal vectors with unit norm and, using the Kronecker δ symbol, are written

$$\mathbf{v}_i \,.\, \mathbf{v}_j = \delta_{ij} \tag{6.11}$$

A set of vectors $\{\mathbf{v}\}$ is *complete* when *any* vector in the space can be expressed as a linear combination of the members of the set $\{\mathbf{v}\}$. In the case of ordinary Euclidean space any group of three of more non-coplanar vectors comprise a complete set. However, it is obvious that a *unique* description of a vector within a given set of base vectors—that is, a unique set of coefficients a_i in a linear expansion,

$$\mathbf{v} = \sum_i^3 a_i \mathbf{v}_i \tag{6.12}$$

is possible only if the base vectors are independent of one another, by which we mean perpendicular or orthogonal. Hence a most important sub-class of complete vectors are defined by *orthonormal bases*. These vector sets have exactly three members in Euclidean space and lie parallel to a set of Cartesian axes. They are often labelled by ($\mathbf{i}$, $\mathbf{j}$, $\mathbf{k}$) but suffix notation makes the labels ($\mathbf{e}_1$, $\mathbf{e}_2$, $\mathbf{e}_3$) more convenient. Altogether, for the orthonormal set $\{\mathbf{e}\}$, we have

$$\mathbf{e}_i \cdot \mathbf{e}_j = \delta_{ij} \tag{6.13}$$

and any vector v may be expanded in terms of these base vectors as in (6.12). The expansion coefficients in (6.12) are easily found by multiplication of both sides of the equation by one member of the base set:

$$\begin{aligned} \mathbf{e}_j.\mathbf{v} &= \mathbf{e}_j\left(\sum_i^3 a_i\mathbf{e}_i\right) \\ &= \sum_i^3 a_i\,(\mathbf{e}_j.\mathbf{e}_i) = \sum_i^3 a_i\delta_{ij} = a_j \end{aligned} \tag{6.14}$$

Hence the expansion coefficients in the linear combination (6.12) are the scalar products $\mathbf{e}_i.\mathbf{v}$ or the projection of $\mathbf{v}$ upon the base vectors.

Of course, any given vector $\mathbf{v}$ may be expressed as a linear combination of three, orthonormal vectors oriented in any way at all. There is an infinite number of such base vector triads, related to one another by rotation. In terms of two bases $\{\mathbf{e}\}$ and $\{\boldsymbol{\epsilon}\}$ we have

$$\mathbf{v} = a_1\mathbf{e}_1 + a_2\mathbf{e}_2 + a_3\mathbf{e}_3 \tag{6.15}$$

and

$$\mathbf{v} \quad = \alpha_1\epsilon_1 + \alpha_2\epsilon_2 + \alpha_3\epsilon_3 \tag{6.16}$$

The expansion coefficients are different in the two cases, of course, even though the vector **v** is the same.

All these properties of Euclidean vectors have close analogies in a different space we call *Hilbert space* named after the mathematician David Hilbert. Already the reader will have observed a number of parallels between the discussions of Euclidean vectors and the introduction of §6.1. Well an explanation of Hilbert space serves to formalize these analogies.

While Euclidean space is three-dimensional, Hilbert space has an infinite number of dimensions. Provided we define a space and its vectors by reference to *mathematical operations* and do not try to visualize these entities, no problem arises. After all, a four-dimensional Euclidean space does not seem 'real' in an everyday sense, but may be constructed by a very obvious extension of what we have just done. In Hilbert space, *scalars* are defined as the set of all *complex numbers* while a *vector* is a *complex function* ψ of a set of real variables $\{x\}$. In this way, a Hilbert-space vector $\psi(x)$ is a *rule of correspondence* that assigns a complex number of each variable set $\{x\}$. Scalar multiplication and vector addition are defined just as in Euclidean space and their combination again describes the concept of the *linear combination of vectors*,

$$\psi = c_1\psi_1 + c_2\psi_2, \tag{6.17}$$

analogous to (6.8).

The *inner-* or *scalar-product* of two Hilbert-space vectors ψ_1 and ψ_2 is defined by

$$\langle\psi_1|\psi_2\rangle \equiv \int \psi_1{}^*\psi_2 \, d\tau, \tag{6.18}$$

where $\int d\tau$ means integral over all variables. In detail; at each value of the variable string $\{x\}$ we multiply ψ_2 by the complex conjugate of ψ_1 and then integrate over all values of the variables. We recognize (6.18) as the *overlap integral* between ψ_1 and ψ_2. Definition (6.18) for the scalar product in Hilbert space genuinely *is* the analogue of (6.9) because of, rather than despite, the inclusion of the integration. This follows because the product $\psi_1{}^*\psi_2$ varies as a *function* of

$\{x\}$ and so does not correspond to a scalar in Hilbert space (which is a complex number): only after integration do we arrive at a quantity—a complex number, in general—which does not depend on the $\{x\}$.

On pursuing the analogy with Euclidean space, the *norm* of a complex function in Hilbert space is just the scalar product of the function with itself,

$$\text{norm } \psi \equiv \langle\psi|\psi\rangle \equiv \int \psi^* \psi d\tau \quad (6.19)$$

and two such functions are orthogonal if

$$\langle\psi_1|\psi_2\rangle = 0. \quad (6.20)$$

Hence we refer to an *orthonormal set of functions* when

$$\langle\psi_i|\psi_j\rangle = \delta_{ij}, \quad (6.21)$$

using the Kronecker function once again.

Referring back to § 6.1 then, the reason for expanding a function (a Hilbert space vector) in terms of an *orthogonal* basis is to provide a *unique* expansion description of the given function in that basis. Had the basis functions not been independent, they could be expanded in terms of one another and the overall expansion would have been non-unique.

We leave this discussion with a table of comparisons between vectors in ordinary space and functions in Hilbert space.

Table 6.1

	Euclidean space	Hilbert space
Scalar	Real number, a	Complex number, a
Vector	Directed line segment, $\mathbf{v}$	Complex function, ψ
Linear combination	$\mathbf{v}=a_1\mathbf{v}_1 + a_2\mathbf{v}_2$	$\psi=c_1\psi_1 + c_2\psi_2$
Scalar product	$\mathbf{v}_1\cdot\mathbf{v}_2$	$\langle\psi_1|\psi_2\rangle$
Norm	$\mathbf{v}\cdot\mathbf{v}$	$\langle\psi|\psi\rangle$
Orthonormal set	$\{\mathbf{e}\}$ with $\mathbf{e}_i\cdot\mathbf{e}_j = \delta_{ij}$	$\{\phi\}$ with $\langle\phi_i|\phi_j\rangle = \delta_{ij}$

6.3 *Crystal-field potentials*

In Chapter 3 we examined the way in which the Coulomb interaction splits up a configuration into terms. The ground term of the d^2 configuration, for example, is ^3F and we refer to the 21 equally favourable electronic arrangements with respect to electronic kinetic energy, attraction to the nucleus, and interelectronic repulsion and exchange. Transition metal ions are characterized by a reasonably well-defined d^n ground configuration. The environment for such ions situated within a crystalline lattice—for example, chromium(III) ions in the aluminium oxide lattice of ruby—ceases to be spherical or isotopic. In that case, the metal electrons—in particular, the *d* electrons—find some positions in space energetically more favourable than others. The electrons must now avoid regions in the crystal with higher electron density as well as avoiding each other. In the limiting case that these effects can be represented by a classical electrostatic field, we talk of the transition metal ion as being subject to a *crystal-field potential, V,* and the Hamiltonian operator for the system is then written

$$\mathcal{H} = \underbrace{\sum_i \mathcal{H}_{\text{H-like}}(i)}_{\text{Configuration}} + \underbrace{\sum_{i<j} \frac{e^2}{r_{ij}}}_{\text{Free-ion terms}} + \underbrace{V}_{\text{Crystal-field terms}} \qquad (6.22)$$

This brief resumé could well serve as an introduction to the subjects known as *crystal-field* and *ligand-field theory* but this is not part of our brief here. However, as a pointer towards such matters and to provide a realistic example of the use of the expansion theorem, the crystal field potential deserves another moment's study. Visualize the variation of this potential in space around the metal ion by the contour map, shown in Figure 6.2.

The crystal-field potential is a continuous function changing throughout the three-dimensional space with a generally complex profile. For many purposes it is convenient to describe it in terms of parts we already know something about; to expand it in terms of a suitable basis. The most commonly used basis—but, remembering

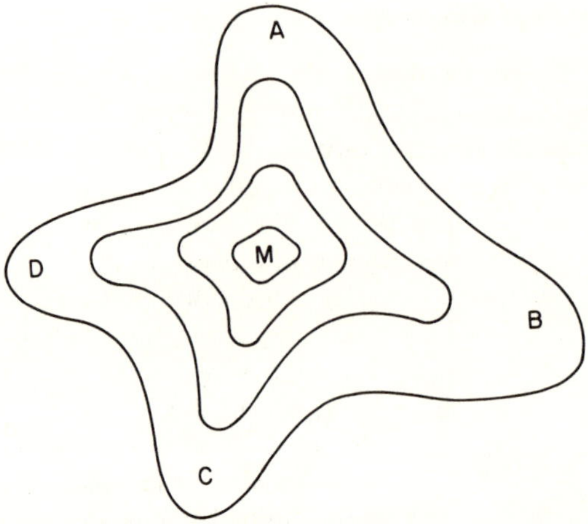

Figure 6.2 The crystal-field potential experienced by the metal *d* electrons in a crystalline environment structured by atoms A, B, C, D.

the arbitrary nature of bases, this is not at all mandatory—is that of the spherical harmonics. These are functions of angular coordinates only, as we know, so that in the expression,

$$V = \sum_i a_i \, Y_m^l(i), \tag{6.23}$$

the coefficients a_i are themselves functions of the third coordinate r. Although we do not investigate this superposition much further here, it is useful to see how the expansion idea can be used: and, in the case of the crystal-field potential, one can imagine without too much difficulty how a contour like that in Figure 6.2 can be built up by superposition of the various *s, p, d, f*... functions, suitably scaled. One final point on this subject: the greater part of textbooks[5–8] on crystal- and ligand-field theory are concerned with highly symmetric metal ion environments, the most common being the octahedron. In these cases, symmetry effectively limits the number of terms that are necessary in an expansion like (6.23). Further, since we are concerned almost exclusively with the potential acting on *d* electrons, a further limitation is placed upon the series expansion. It is the case, for example, that the crystal-field potential acting on *d* electrons in a perfectly octahedral environment is resentable by the very simple and complete, expression,

$$V_{\text{oct}} = aY_0^0 + b\,[Y_0^4 + \sqrt{\tfrac{5}{14}}(Y_4^4 + Y_{-4}^4)]. \tag{6.24}$$

6.4 *Perturbation Theory*

Although we do not carry out any explicit energy computations in this book, it is useful to learn something of how these might be done. It is a fact of life that all such calculations are inexact and that approximations of one kind or another just have to be made in practice. Perturbation methods comprise one group of approximation techniques widely employed in computational quantum chemistry. The following review seeks merely to summarize the philosophy and utility of the approach.

Faced with the problem of solving the Schrödinger equation

$$H\psi = E\psi \tag{6.25}$$

for some system of interest, the perturbation method begins by looking for another problem

$$H^0\psi^0 = E^0\psi^0, \tag{6.26}$$

that has already been solved and which bears some reasonably close relationship to the task at hand. By this is meant that if we write

$$H = H^0 + H^{(1)}, \tag{6.27}$$

the additional term—or perturbation—$H^{(1)}$ gives rise to sufficiently small changes in energies and eigenfunctions that the subsequent series expansion we shall discuss will converge sufficiently quickly.

Without considering the derivation of the promised series expansions in any detail here, we can make one observation upon the strategy it employs. In seeking solutions to (6.25), we recognize in any of the eigensolutions an unknown function of some fixed set of variables with some unknown, but no doubt well-behaved, boundary properties. We seek, therefore, to expand the unknown ψ in terms of a basis of known functions of the same variables and with similar boundary properties. What could be more convenient here than the solutions to the 'closely similar' problem, described by (6.26)? In short, we seek to expand the functions after perturbation as superpositions of those before perturbation. This is the key idea in the perturbation method.

After some not very difficult algebra, the following expressions are produced by so-called 'non-degenerate' perturbation theory:

$$E_n = E_n{}^0 + H_{nn}^{(1)} + \sum_m{}' \frac{H_{nm}^{(1)} H_{mn}^{(1)}}{E_n{}^0 - E_m{}^0} + \ldots \tag{6.28}$$

$$\psi_n = \psi_n{}^0 + \sum_m{}' \left(\frac{H_{nm}^{(1)}}{E_n{}^0 - E_m{}^0} \right) \psi_m{}^0 + \ldots \tag{6.29}$$

Their significance is explained against the pictorial summary in Figure 6.3. On the left are represented the eigenvalues and eigenvectors of $\mathcal{H}^0$: everything here, marked with a right zero superscript, is presumed known. The eigenvalues and vectors on the right are those we seek and are solutions of $\mathcal{H}$ which is also known precisely: if it were not, we would not even know what problem we were trying to solve! Now consider the nomenclature in (6.28) and (6.29). The $H_{ij}^{(1)}$ are integrals of the form

$$H_{ij}^{(1)} \equiv \langle \psi_i{}^0 | \mathcal{H}^{(1)} | \psi_j{}^0 \rangle. \tag{6.30}$$

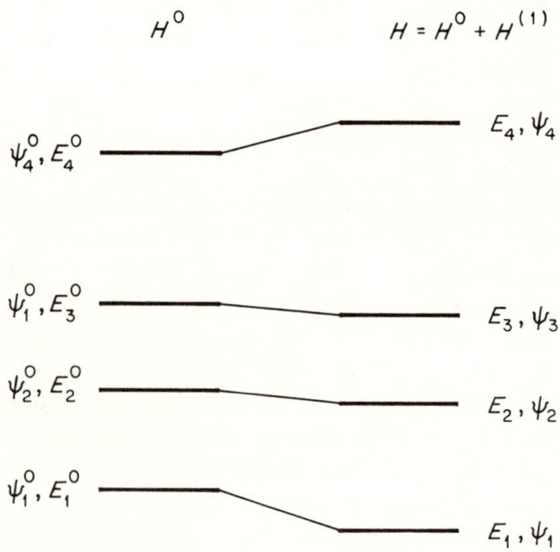

Figure 6.3 The 'old' energies E_i^0 and wavefunctions ψ_i^0 are perturbed by $\mathcal{H}^{(1)}$ to give the 'new' energies E_i and wavefunctions ψ_i.

They are frequently called *matrix elements*; or, more fully here, matrix elements of the perturbation Hamiltonian, $H^{(1)}$, within the basis $\{\psi^0\}$. The reason for the name is subtle if one delves deeply into matrix mechanics; but obvious if one simply recognizes a matrix as a table. Thus, we can construct all integrals under a given operator—here $H^{(1)}$—between all possible members of a chosen basis—here $\{\psi^{\circ}\}$—as in Figure 6.4. A matrix element is then just an element of such a matrix!

As $H^{(1)}$ is part of a Hamiltonian (energy) operator, the matrix elements in (6.30) have units of energy. We can talk of diagonal, $H_{ii}^{(1)}$, and non-diagonal elements, $H_{ij}^{(1)}$, but they are all energies. Therefore, all terms in (6.24) have dimensions of energy, as they should; and all terms in (6.29) are dimensionless. Next, observe how *all* quantities on the right of either (6.28) or (6.29) involve *only*

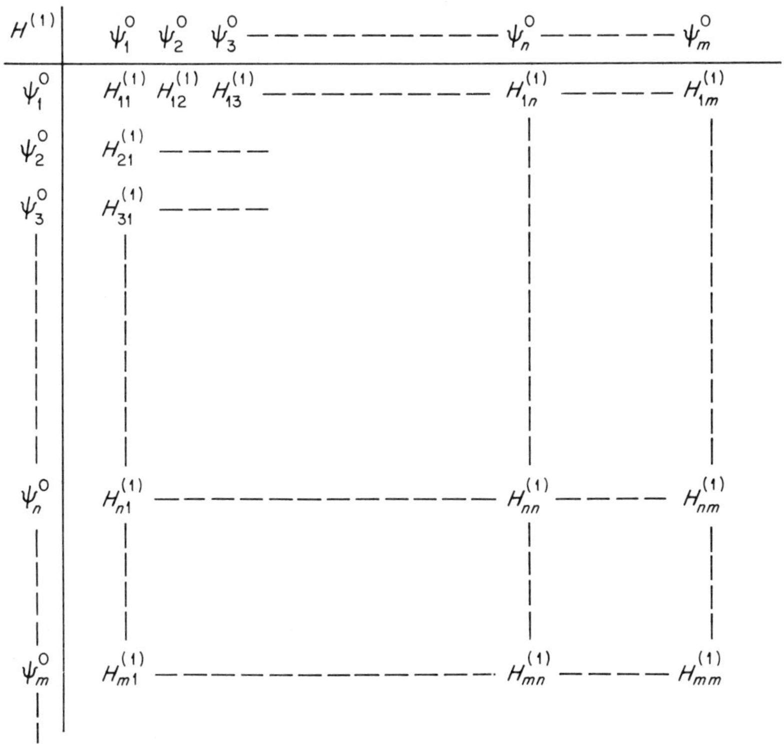

Figure 6.4 The matrix of $H^{(1)}$ in the basis $\{\psi\}$.

known parts—$\{E^0\}$, $\{\psi^0\}$, or $\mathcal{H}^{(1)}$—and so these perturbation expansions do indeed furnish means for expressing the unknown energies $\{E\}$ and wavefunctions $\{\psi\}$ in terms only of known quantities. The prime on the summation signs in these equations signifies that the sum is to run over all members of the known basis *except* for the case $m = n$.

Now consider the expression in (6.29) for the wave*functions* we seek. If we retain only the first term on the right, we have a so-called zeroth-order description; namely that the required function on the right of Figure 6.3 is the same as that on the left. The second term, which is called the first-order term, generally involves a whole series of parts as implied by the summation sign. The sum is over all members of the set $\{\psi^0\}$, other than the ψ_n^0 under consideration. The expansion coefficients are dimensionless quantities, as discussed, and hence just numbers: in general, they may be complex numbers. For concreteness, let us invent some coefficients (which we take as real, for simplicity) and write the first-order perturbation expression for the second eigenfunction, say, as

$$\psi_2 = \psi_2 + 0.1\psi_1{}^0 - 0.02\psi_3{}^0 + 0.06\psi_4{}^0 + \ldots \qquad (6.31)$$

Written out like this, we see that (6.29) simply expresses the unknown functions as superpositions of the known.

A lot of jargon attaches to this subject and it is as well to get acquainted with it from the beginning. Since the 'new' wavefunction is expressed as the 'old' plus varying amounts of the rest of the basis $\{\psi^0\}$, we say that the perturbation $\mathcal{H}^{(1)}$ *mixes in* the various components of the basis into the function under consideration: explicitly, $\mathcal{H}^{(1)}$ mixes excited functions $\{\psi^0\}$ into the ground function $\psi_1{}^0$, for example. However, no particular reality can be attached to this notion, for the choice of basis is merely a matter of convenience. A more correct, if pedantic, statement would be that the new function looks 'as if' such-and-such basis functions had been admixed. Now the *mixing coefficients* $[H_{nm}^{(1)}/(E_n^0 - E_m^0)]$ are inversely proportional to the energy separation between the functions being mixed (*before* mixing) and proportional to the matrix element $H_{nm}^{(1)}$ that *connects* the two functions. Sometimes, different basis functions can be characterized by labels (symmetry labels) that allow us to recognize at a glance that a given perturbation $\mathcal{H}^{(1)}$ cannot connect them. Then the numerator is zero and the functions do not mix. As a great deal of time and effort can be saved by recognizing such null effects in advance, appropriate labelling schemes are frequently very important in practice. Although we shall note a few examples in due course, it is not part of the theme of this book to explore this aspect of the subject further, however.

Turning now to the expression (6.28) for the new eigen*values*, we note first that the first-order term $H_{nn}^{(1)}$ involves a single, diagonal matrix element only. This means that, to this level of approximation, the expression for the new energy is independent of the rest of the basis. The first order 'correction' can be positive or negative but must be *real*. This last point arises because of the property, of all quantum mechanical operators associated with the observables through an eigenvalue equation, that we call hermiticity. *Hermitian operators*, by definition, satisfy the relationship.

$$\int \phi^* \hat{O} \psi \, d\tau = \int \psi \hat{O}^* \phi^* \, d\tau \tag{6.32}$$

or

$$\langle \phi | \hat{O} | \psi \rangle = \langle \psi | \hat{O} | \phi \rangle^*. \tag{6.33}$$

It follows that a matrix like that in Figure 6.4 is Hermitian: that is, one whose elements related by reflection in the leading diagonal are complex conjugates of one another. On the diagonal itself, of course, that relationship—which is just (6.33)—can only be satisfied if the elements are real. The ultimate reason why such operators are constructed to be Hermitian is to ensure that the observables are real: a complex or imaginary observable would be a contradiction. One example of making an operator Hermitian that we have already studied is the appearance of the imaginary i in the angular momentum operator, $i\partial/\partial x$: see also §7.5. In the second-order term of (6.28) the numerators in the sum are real because a complex number times its complex conjugate is real and so all parts of the exprêssion for the observable energy are then real, as they should be. In the next section we study the behaviour of (6.28) in a little more detail and so provide one view of the 'non-crossing' rule that we used in Chapter 4.

In tying up loose ends, the 'non-degenerate' perturbation theory discussed here refers to the situation in which the solutions of the known problem (6.26) are non-degenerate. Often that does not obtain. Rather similar expressions to those in (6.28) and (6.29) are derived in the so-called 'degenerate' theory but that is a matter we do not pursue in this book.

6.5 *The non-crossing rule (I)*

Suppose, only for simplicity, that we are concerned with a system possessing only two eigensolutions. Figure 6.5 is constructed appropriately for this situation and in the manner of Figure 6.3, except that we assume that the first-order energy shifts of the two levels are accidentally zero. This is merely a stratagem allowing us to focus on the second-order term in (6.28). The numerators in these terms are positive, because $H_{21}^{(1)} = H_{12}^{(1)*}$ by hermiticity, and are clearly equal for the two levels. The denominators are equal also, at least in magnitude. However, they differ in sign and are such that the lower level suffers a downward shift in energy while the upper one shifts upwards. Indeed, quite generally, regardless of the number of levels concerned, the ground function must decrease as E_1^0 is, by definition, the lowest energy; and by the same token, the highest function must shift upward. Levels in between may go either way depending upon the sum in (6.28).

Now let us introduce first-order perturbation shifts into the picture and suppose, as is not uncommon but by no means necessary, that these raise the energy of the lowest level and lower that of the upper one. The first-order situation is represented in Figure 6.6 by the solid line. These energy levels change linearly because first-order energy shifts take the form, from (6.28), of $H_{nn}^{(1)}$ which are

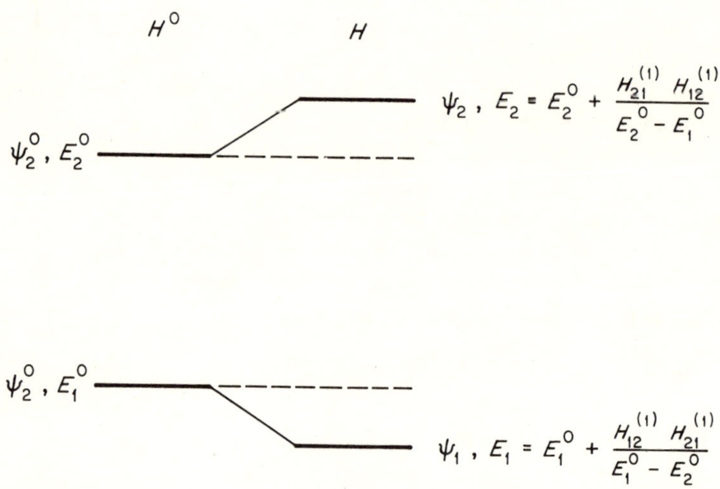

Figure 6.5 The second-order perturbation term describes how the lower energy wavefunction is lowered and the higher energy one raised by the same amount.

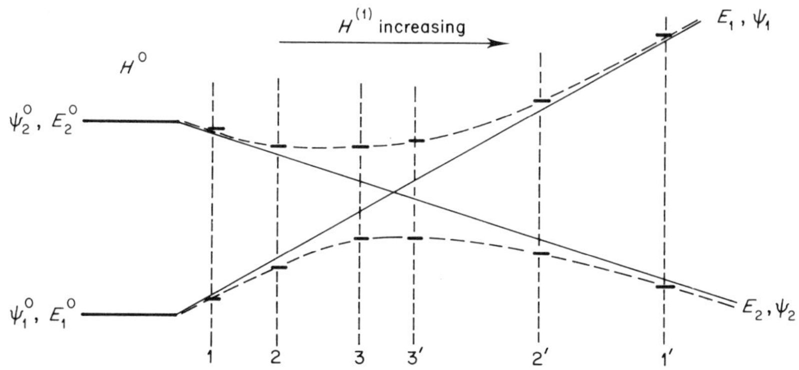

Figure 6.6 Illustrating the non-crossing rule by perturbation theory. Labelling of the functions on the right is paradoxical but correct.

obviously proportional to the operator $H^{(1)}$ from the definition (6.30). Because of our particular choice, these levels are predicted to cross at some value of $H^{(1)}$. However, it is quite incorrect to talk of first- and second-order perturbation *effects*. There is only the one perturbation: it is just that our theory chooses to represent the results as series expansions—once again, merely a matter of convenience—and we have labelled the terms in those expansions as of first- and second-order. The mathematics is first and second order, not the physics. So we are *obliged* to consider a second (and strictly higher) order correction concurrently with the first-order ones. These are given by the expressions in Figure 6.5, as we have seen. Since the denominators are fixed (involving energies with zero superscripts), these corrections are quadratic in $H^{(1)}$ and, as we have discussed, such as to cause energy shifts to higher and lower energies for the upper and lower levels, respectively. It is as if the two levels *repel* one another. The reader will note that the dotted lines, representing the energies of the two levels after all terms of the perturbation have been recognized, tend asymptotically towards the limiting first-order slopes. To see why, imagine that we perform a number of separate perturbation calculations on the system starting at physical situations represented by the vertical lines 1, 2, 3. Assume that the starting eigensolutions (those with the zero superscripts) at each point are taken as those obtained by an earlier calculation: the appropriate energies at each point are represented in the figure by the short horizontal lines. Now for each calculation, the first-order terms will become numerically progressively smaller. That is because the new zeroth-order wavefunctions progressionally involve an increasingly equal mixture of the original functions ψ_1^0 and ψ_2^0 and so the net first-order terms tend towards equal and average values. At the same time, the appropriate energy denominators of the second-order term decrease as we approach the notional crossing point and so these terms begin to dominate the scene. As we can similarly consider approaching the crossing point from the right, through the sequence 1′, 2′, 3′, we see that the maximum 'repulsion' between the two levels occurs at the notional crossing point. *Provided* that the original functions ψ_1^0 and ψ_2^0 are indeed mixed by the perturbation Hamiltonian in question—that is, provided $H^{(1)}$ connects ψ_1^0 and ψ_2^0—such levels cannot cross. Note also that, for values of $H^{(1)}$ far removed from the non-crossing point (as we may now call it), the intermixing of the original functions ψ_1^0 and ψ_2^0 is minimal because of the large energy denominator in the first-order term of (6.29), so it is sensible to label the wavefunctions as shown in the figure. Near the non-crossing point, such mixing becomes intimate and the labels ψ_1 and ψ_2 (without superscripts) can be misleading. Overall, however, because the energies follow the dotted lines in the figure, we observe ψ_2^0 becoming ψ_1 and ψ_1^0 becoming ψ_2. It is as if the levels have crossed but without actually doing so! Any confusion, however, only relates to the labelling scheme and not to the actual results.

We leave the non-crossing rule for the moment but consider a different derivation of it shortly.

6.6 *The variation method*

The perturbation method works best when $H^{(1)}$ is small; in effect, when the perturbation series expansions (6.28) and (6.29) converge rapidly. If the approximation afforded by the first-, or perhaps second-, order terms are insufficiently exact for a given purpose, the tedium of computing third- and higher-order approximations—which are, of course, extensive series in themselves—will usually not be worth the effort. The variation method we now review will often provide a better way: indeed, the perturbation technique can be shown to be a kind of subdivision of the variation method, though we shall not pursue this matter here.

The variational principle states that the expectation energy of any system wavefunction we may care to propose will be greater than, or just possibly equal to, the true ground state energy of the system. It provides a sense of direction for refinement or variation of a guess. The proof goes as follows. Let the true eigensolutions, $\{\psi\}$ and $\{E\}$, satisfy the Hamiltonian operator $\mathcal{H}$ of the system at hand:

$$\mathcal{H}\psi_n = E_n\psi_n. \tag{6.34}$$

In particular, let the ground state solution be

$$\mathcal{H}\psi_0 = E_0\psi_0 \tag{6.35}$$

Suppose, not knowing the true ground state wavefunction for our system, we propose—by insight, guesswork or blind inspiration—the function ϕ for this role. We may sensibly consider the integral

$$E' = \langle\phi|\ \mathcal{H}\ |\phi\rangle \tag{6.36}$$

for if our guess had been exactly right (and assuming ϕ to be normalized), this is just the true ground state energy. If, as is almost certain, our guess is not right we enquire what E' means. To do this we notionally expand our proposed *variation function* ϕ in terms of the complete set of orthonormal, true eigenfunctions $\{\psi\}$ of the system

$$\phi = \sum_i a_i\psi_i; \qquad \sum_i a_i{}^*a_i = 1 \tag{6.37}$$

On substitution into (6.36), we find

$$\begin{aligned} E' &= \sum_i\sum_j a_i{}^*a_j\ \langle\psi_i|\ \mathcal{H}\ |\psi_j\rangle \\ &= \sum_i a_i{}^*a_iE_i. \end{aligned} \tag{6.38}$$

The last line follows, because the $\{\psi\}$ form an orthonormal set of eigenfunctions of $\mathcal{H}$ as in (6.34): thus

$$\langle\psi_i|\ \mathcal{H}\ |\psi_j\rangle = \langle\psi_i|E_j\psi_j\rangle = E_j\langle\psi_i|\psi_j\rangle = E_i\delta_{ij} \tag{6.39}$$

Then, subtracting E_0, the lowest eigenvalue of the system, from both sides of (6.38), we find

$$E' - E_0 = \sum_i a_i^* a_i (E_i - E_0), \tag{6.40}$$

where we have used the second relationship in (6.37), so that

$$\sum_i a_i^* a_i E_0 \equiv E_0 \sum_i a_i^* a_i = E_0. \tag{6.41}$$

Now, since all E_i in (6.40) are greater than or equal to E_0, by definition (6.35), and the coefficient products $a_i^* a_i$ are non-negative real numbers, it follows that

$$E' - E_0 \geqslant 0. \tag{6.42}$$

That is, the expectation energy of any guessed variation function ϕ is greater than, or equal to, the true ground state energy: this is the variational principle. The idea can be extended, with some care, to the computation of excited states also, but we shall not consider those cases. We look only at the use of the principle to optimize approximations to the *ground* state energy of a system.

Application of the principle can take many forms but all have one central theme. If we make a series of guesses, systematically or not, at the ground state wavefunction of the system, the variational principle tells us that that guess associated with the lowest energy is the best of the group. If we do set up a systematic variation of the trial wavefunction, the principle provides a sense of direction for its refinement. One important and very common means of exploring the method is the Rayleigh–Ritz technique that we now investigate.

6.7 *Molecular orbitals*

In the following chapter we shall discuss labels for the orbitals and terms of linear molecules. We can prepare for that and illustrate the Rayleigh–Ritz variation method at the same time. We consider the bonding between two atoms, 1 and 2, and for present purposes it is sufficient to limit our concern to the case of so-called σ bonding. A thorough description of such labels appears in the next chapter but the idea will be well enough known at this stage.

The molecular Hamiltonian will comprise the kinetic energy, Laplacian operator as usual; the Coulomb operator to express electron–electron interaction; and also the binuclear terms of nuclear–nuclear repulsion and electron–nuclear attraction—not only for an electron to its own nucleus but also to the other:

$$H = -\frac{1}{2m}\sum_i \nabla^2(i) + \frac{Z_1 Z_2 e^2}{R} - \sum_i \frac{Z_1 e^2}{r_1(i)} - \sum_i \frac{Z_2 e^2}{r_2(i)} + \sum_{i<j} \frac{e^2}{r_{ij}}. \tag{6.43}$$

Whatever the ground orbital looks like it must be of some form encompassing the space around both nuclei for it describes, by assumption here, a bonding situation: the nuclei are held together and the valence electrons are associated with the whole molecule. We can guess at some wavefunction like that in Figure 6.7. For

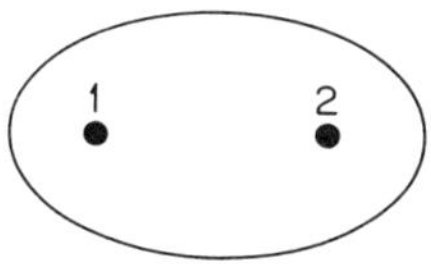

Figure 6.7 A molecular orbital resembles atomic orbitals near the nuclei but not elsewhere.

reasons that should be obvious, but are discussed at length in the next chapter anyway, the orbital will be cylindrically symmetric with respect to the intermolecular axis. Putting aside anything we may have read elsewhere, that is almost as much as we can say at this point. The only other useful observation is that *near* the nuclei the molecular wavefunction must look rather like appropriate atomic wavefunctions, for the dominant influences upon the electrons in such vicinities are just those of the local atom: molecular effects are strongly overriden. But in the region between the nuclei we cannot really say anything very useful yet.

So here we have an unknown function of a set of variables with boundary properties not unlike those of free atoms. Why not choose to expand it in terms of atomic functions, for we know about those and we will simultaneously get the variable list and other conditions right straight away. We construct an expansion, then, for a molecular orbital in terms of a linear combination of atomic orbitals—LCAO–MO:

$$\psi = c_1 1s_1 + c_1' 1s_2 + c_2 2s_1 + c_2' 2s_2 + c_3 2p_{x1} + c_3' 2p_{x2} + \ldots + 3d \ldots + 4f \ldots \text{ etc.} \qquad (6.44)$$

Once again, the choice of this kind of superposition is free, has no intrinsic physical significance, and is made for reasons of convenience—and the latter usually means, because the series expansion can be curtailed fairly soon. Also, once again, the series is infinitely long in principle. But now we make an enormous simplifying assumption; that we can obtain sufficiently satisfactory results with only a few leading terms—in fact with only two! For the H_2 molecule we just use the 1*s* atomic orbitals on each atom. For molecules like CO we retain only one so-called 'valence orbital' on each atom—here C2*s* and O2*s*, say. It is common to see this sort of simplification justified by recourse to the point that the molecular orbital will closely resemble these atomic orbitals near the two nuclei: quite so, but that cannot be expected to be satisfactory elsewhere in the orbital. However, great insight into bonding can be had from such simple beginnings, so the idea is useful: we must remember, however, not to expect too much from so restrictive a simplification.

So let us agree to work with an expansion of the form

$$\psi = c_1\chi_1 + c_2\chi_2 \qquad (6.45)$$

where χ_1 and χ_2 are valence atomic orbitals on centres 1 and 2 and ψ is the molecular orbital we seek to describe. We are free to vary the 'mixing coefficients' c_1 and c_2, though we shall make ψ normalized by the condition

$$c_1^2 + c_2^2 = 1: \qquad (6.46)$$

in this section we can take all quantities to be real, rather than complex, without any essential loss of generality. Our task now is to optimize values for these coefficients and we use the variational principle to do it.

6.8 *The secular equations*

The electronic energy associated with the wavefunction ψ and Hamiltonian H is

$$E = \langle\psi|\, H\, |\psi\rangle / \langle\psi|\psi\rangle \tag{6.47}$$

or, when substituted by the LCAO expansion (6.45),

$$E = \frac{\langle c_1\chi_1 + c_2\chi_2|H|c_1\chi_1 + c_2\chi_2\rangle}{\langle c_1\chi_1 + c_2\chi_2|c_1\chi_1 + c_2\chi_2\rangle} \tag{6.48}$$

Writing,

$$H_{ij} \equiv \langle\chi_i|H|\chi_j\rangle = \langle\chi_j|H|\chi_i\rangle \equiv H_{ji} \tag{6.49}$$

and

$$S_{ij} \equiv \langle\chi_i|\chi_j\rangle = \langle\chi_j|\chi_i\rangle \equiv S_{ji}, \tag{6.50}$$

we transcribe (6.48) with simplified nomenclature, as

$$E = \frac{c_1^2 H_{11} + 2c_1c_2H_{12} + c_2^2H_{22}}{c_1^2S_{11} + 2c_1c_2S_{12} + c_2^2S_{22}} \tag{6.51}$$

Now the variational principle tells us that the best molecular orbital we can construct from the basis we have chosen to restrict as χ_1 and χ_2 in (6.45) is that yielding the minimum energy. Therefore we seek the conditions under which $\partial E/\partial c_1=0$ and $\partial E/\partial c_2=0$. In differentiating (6.51), a useful trick that simplifies the algebra is to consider the expression as a product uv, where u is the numerator of (6.51) and v is the reciprocal of the denominator:

$$\frac{\partial E}{\partial c_1} = \frac{(2c_1H_{11} + 2c_2H_{12})}{(c_1^2S_{11} + 2c_1c_2S_{12} + c_2^2S_{22})}$$

$$- \frac{(2c_1S_{11} + 2c_2S_{12})(c_1^2H_{11} + 2c_1c_2H_{12} + c_2^2H_{22})}{(c_1^2S_{11} + 2c_1c_2S_{12} + c_2^2S_{22})^2} = 0 \tag{6.52}$$

for minimum energy. Multiply through by $\tfrac{1}{2}(c_1^2S_{11} + 2c_1c_2S_{12} + c_2{}^2S_{12})$ and rearrange:

$$c_1H_{11} + c_2H_{12}$$

$$- (c_1S_{11}+c_2S_{12})\left(\frac{c_1^2H_{11} + 2c_1c_2H_{12} + c_2^2H_{22}}{c_1^2S_{11} + 2c_1c_2S_{12} + c_2^2S_{22}}\right) = 0 \qquad (6.53)$$

and, recognizing the last bracket as E from (6.48), we get

$$c_1H_{11} + c_2H_{12} - E(c_1S_{11} + c_2S_{12}) = 0$$

or

$$(H_{11} - ES_{11})\, c_1 + (H_{12} - ES_{12})c_2 = 0 \qquad (6.54)$$

Similarly we can derive, from $\partial E/\partial c_2$, that

$$(H_{21} - ES_{21})c_1 + (H_{22} - ES_{22})c_2 = 0. \qquad (6.55)$$

The pair of equations (6.54) and (6.55) are known as the *secular equations* for this problem. The word 'secular' does not refer to any non-religious aspect of the subject but rather to the fact that the classical version of variational and perturbation theories were often used to compute the periods of planetary motions!

Our secular equations are a simultaneous pair in the unknowns, c_1 and c_2 whose values for minimum energy we seek. However, that value of energy appearing in these equations is also unknown to us. A well-known theorem of simultaneous linear equations tells us that their independence (that is, that (6.54) and (6.55) are indeed different) is only guaranteed if the determinant of the coefficient of the variables (c_1 and c_2) vanishes: that is, if

$$\begin{vmatrix} H_{11}-ES_{11} & H_{12}-ES_{12} \\ H_{21}-ES_{21} & H_{22}-ES_{22} \end{vmatrix} =0 \qquad (6.56)$$

This determinant is called the *secular determinant,* and (6.55) yields a quadratic equation in E,

$$(H_{11}-ES_{11})(H_{22}-ES_{22}) - (H_{12}-ES_{12})(H_{21}-ES_{21}) = 0, \qquad (6.57)$$

with two roots, say E_1 and E_2. We can now substitute these energies, one at a time, back into the secular equations (6.54), (6.55) For example.

$$(H_{11} - E_1S_{11})c_1 + (H_{12}-E_1S_{12})c_2 = 0 \qquad (6.58)$$

in which only c_1 and c_2 remain as unknowns: therefore we obtain a ratio for c_1/c_2. We cannot obtain the c values explicitly by substituting E_1 into (6.55) as well, for it transpires that we arrive at an identical equation to (6.58). Instead, we use the normalization condition (6.46) to arrive finally at c_1 and c_2. We can then repeat the whole process by substituting E_2 into either (6.54) or (6.55) and obtain another pair of c values. Altogether we find two eigensolutions:

$$E_1: \qquad \psi_1 = c_{11}\chi_1 + c_{12}\chi_2 \qquad (6.59)$$
$$E_2: \qquad \psi_2 = c_{21}\chi_1 + c_{22}\chi_2 \qquad (6.60)$$

where we have added a subscript to the coefficients so as to associate a given wavefunction ψ_i with its energy E_i.

The wavefunction with the lesser energy is the best that can be done within the chosen basis (here χ_1 and χ_2) in approaching the ground orbital of the system specified by $\mathcal{H}$. That with the higher energy is merely related to it by orthogonality and is often called a *virtual orbital*. Had we been prepared to include more terms in the LCAO–MO expansion—to use a larger basis—then a similar procedure would have involved equating n differentials $\{\partial E/\partial c\}$ to zero and so ending up with n simultaneous secular equations in n unknowns. The secular determinant would have been $n \times n$ in size, and the determinantal equation,

$$\begin{vmatrix} H_{11}-ES_{11} & H_{12}-ES_{12} & \cdots & H_{1n}-ES_{1n} \\ H_{21}-ES_{21} & .:\ldots & \vdots & \\ \vdots & \cdot & & \cdot \\ H_{n1}-ES_{n1} & \ldots\ldots & \ldots\ldots & H_{nn}-ES_{nn} \end{vmatrix} = 0, \qquad (6.61)$$

would yield n energies $\{E\}$. On back-substitution into the secular equations, n sets of coefficients would be found: $n-1$ of these would describe virtual orbitals and the nth, of lowest energy, would represent the best that could be done with the n-fold basis in approaching the true ground orbital.

6.9 *Diagonalization*

Consider the quadratic equation (6.57) which led to the roots E_1 and E_2. Knowing these roots, we could construct another (and unique) quadratic equation with the same roots that can be solved by inspection; namely,

$$(E_1 - E)\,(E_2 - E) = 0 \tag{6.62}$$

Just as (6.57) arose by expansion of the determinantal equations (6.56), this new quadratic may be considered to have arisen from the determinantal equation

$$\begin{vmatrix} E_1 - E & 0 \\ 0 & E_2 - E \end{vmatrix} = 0 \tag{6.63}$$

whose form is diagonal. This same equation may be arrived at by using the known solutions E_1 and E_2 in a different way, as follows.

We have seen how E_1 and E_2 are associated with the wavefunctions ψ_1 and ψ_2 by (6.59), (6.60). These functions are just linear combinations of our base functions χ_1 and χ_2. Suppose, by lucky guesswork (or hindsight!) we had chosen ψ_1 and ψ_2 as our bases for $\mathcal{H}$. Corresponding to these vectors in Hilbert space, this simply means that we imagine a pair of base vectors rotated with respect to the original pair. So by gazing into a crystal ball, it is conceivable that we could have chosen $\{\psi\}$ to begin with. In that case, the secular determinantal equation we should have produced instead of (6.56) would take the form

$$\begin{vmatrix} H'_{11} - ES'_{11} & 0 \\ 0 & H'_{22} - ES'_{22} \end{vmatrix} = 0 \tag{6.64}$$

because this, being of identical form to (6.62) with an orthonormal basis (so that $S'_{11} = S'_{22} = 1$), yields the roots $E_1 = H'_{11}$; $E_2 = H'_{22}$. As the diagonal matrix blocks off, we have two equations

$$(H'_{11} - ES'_{11})\, c_1 = 0$$
$$(H'_{22} - ES'_{22})\, c_2 = 0 \qquad (6.65)$$

so that root $E_1 = H'_{11}$ is associated with c_1 only, that is with the basis function ψ_1, while E_2 is associated only with ψ_2. All this, of course, is just cyclically consistent with (6.59) and (6.60). Drawing a parallel with Euclidean-space vectors again, the choice of $\{\psi\}$ as base vectors means that the 'unknowns', ψ_1 and ψ_2 are each uniquely expressible in terms of each of the base vectors: the projections of each ψ onto the base vectors are zero or unity; there are no off-diagonal terms.

All of this discussion can be generalized in a formal way and we refer to the processes, by which an nth-tic determinantal equation like (6.61) is solved to produce n eigenvalues together with their associated eigenvectors, as *diagonalization*. It is the same process as one which, by suitable matrix reorganization, transforms the secular determinant into diagonal form: and, in that form, the determinant is *unique*.

While one purpose of presenting this section has been to introduce the jargon, it must not be forgotten that what lies behind it all is the variational principle and the desire to optimize a guess at the ground state wavefunction. The processes have been illustrated by reference to the construction of molecular orbitals, but the ideas carry over to the study of molecular states, involving antisymmetrized products of orbitals, as well. In those cases where we seek optimal combinations of functions on one centre only, or in any other system where the basis functions can be formed into an orthonormal set (for this was not necessary, nor appropriate, in the treatment above), the *overlap matrix*, comprising elements S_{ij}, is just a unit matrix and all mention of S can be dropped from (6.61): problems in ligand-field theory, for example, make use of this simplification. We shall discuss another simple case in the next chapter.

6.10 *The non-crossing rule (II)*

Consider again the quadratic equation (6.57) but in circumstances where the basis functions are normalized:

$$(H_{11}-E)(H_{22}-E) - (H_{12}-ES_{12})^2 = 0 \qquad (6.66)$$

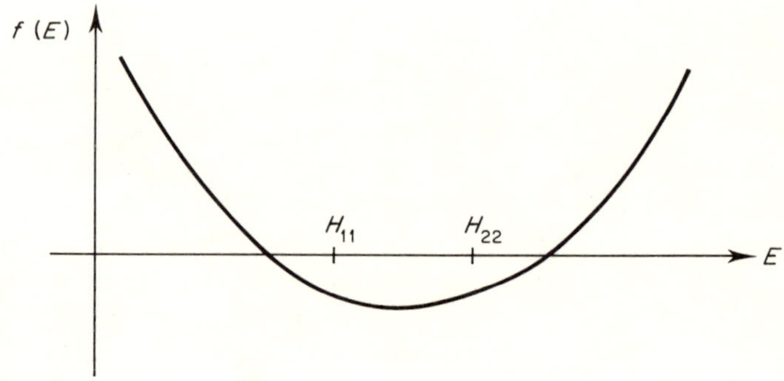

Figure 6.8 Illustrating the non-crossing rule by the variational principle.

Let us assume, as we may just by relabelling, that $H_{11} < H_{22}$. Writing the lefthand side as $f(E)$, we plot F as a function of E, as in Figure 6.8. The curve traces a parabola with $f(E)$ being positive for large E values (positive or negative); and negative when $E = H_{11}$ or H_{22}. These statements are true regardless of the relative magnitudes of H_{11}, H_{22}, H_{12} and S_{12}. Hence we find that the solutions, $E = E_1$ or E_2, corresponding to $f(E) = 0$ as in (6.66) must occur *outside* the bounds H_{11} and H_{22}. It is as if the basis functions, with the diagonal energies H_{11} and H_{22}, had repelled one another and we see the 'non-crossing' rule once again. Of course, if it were inappropriate to combine χ_1 and χ_2 to give ψ_1 and ψ_2, as in (6.45), the 'repulsion' would not occur. Rules of symmetry can reveal such circumstances. The non-mixing corresponds to there being no matrix element $\langle\chi_1|\mathcal{H}|\chi_2\rangle$ between the basis functions; and we recall our earlier remarks towards the end of §6.5.

CHAPTER 7

Diatomic Molecular Orbitals

7.1 *Conserved angular momentum in linear molecules*

In moving on to these simplest of all molecules, we maintain the spirit of this book and confine our attention to the *classification* of orbitals and many-electron wavefunctions. From the outset we exploit a central aspect of all molecules that arises from the greatly dissimilar masses of electrons and nuclei. Velocities of the nuclei within a molecule are small compared with those of the electrons and we may usually regard the motion of the electrons as taking place relative to essentially fixed nuclei. Such is the subject matter of the so-called *Born–Oppenheimer approximation*—a somewhat subtle matter that we do not develop here. Suffice it to note that by referring to the fixed-nucleus approximation we seek to describe the electronic states of diatomic molecules, characterized by the inter-nuclear vector, as possessing axial or cylindrical symmetry. We begin by enquiring how angular momentum descriptions change on replacing a spherical atom by a cylindrical molecule.

The angular momentum commutation rules (2.12) are intrinsic to the nature of angular momentum and the isotropy of free space: in this sense they are always 'correct'. What is important physically, however, is whether they are *relevant*; that is, whether they relate to observable, conserved quantities in any given system. For a spherical atom, the only term in the Hamiltonian involving the polar angle ϕ arises from the Laplacian as $\partial^2/\partial\phi^2$. This obviously commutes with the $\partial/\partial\phi$ appearing in the polar form of the operator l_z and so H and l_z commute. As atoms are spherical, all directions are equivalent ensuring, despite the seemingly different shape of the relevant mathematical expressions, that H also commutes with components of angular momentum referred to *any* direction whatever—and specifically H commutes with l_x and l_y. The total angular momentum operator squared, given by the sum of, say, the cartesian components squared, therefore commutes with H also. So, each of l_x, l_y, l_z, l^2 commutes with the central-field Hamiltonian, although we know from (2.12) that the components do not commute amongst themselves. In choosing arbitrarily, but without loss of generality, z as the quantization axis, we settled on the commuting operator set H, l^2 and l_z. Then, exploiting the theorem (§ 2.3) about commuting operators sharing a common set of eigenfunctions, we were able to conclude that definite values for the observables associated with these operators could be determined simultaneously by appropriate experiments. It is true that H commutes with all the aforementioned operators so H and l_x, for example, also share a common set of eigenfunctions and definite values for their associated properties can be measured simultaneously. But, as l_x and l_z do not commute, the common eigenfunction set for H and l_x cannot be the same as that for H and l_z. In short, the best that we can do, in defining conserved properties and constants of the motion, exploits the Hamiltonian, the total orbital angular momentum, and one (freely selected) component of it. All this aside from spin, of course . . .. Nevertheless the commutation of the spherical Hamiltonian with l_x and l_z was essential to establish the same property for H and l^2.

Against this review for atoms, consider what can be said of angular momentum in the cylindrical symmetry of linear, molecules. Operators l^2, l_x, l_y, l_z exist as before and their commutation relationships remain as given in (2.12): but they mostly cease to have relevance in the linear system. This follows because only l_z—taking z to lie along the molecular axis—commutes with the cylindrical Hamiltonian. That l_z *does* commute with the molecular Hamiltonian follows immediately from the fact that a view along that axis fails to distinguish a linear molecule from an atom: the ϕ angle, taken to mean angular displacement about the *z*-axis, in parallel with the polar case, again only enters into the molecular Hamiltonian in the Laplacian ∇^2 as $\partial^2/\partial\phi^2$ and this commutes, as before, with the $\partial/\partial\phi$ of l_z. The electronic angular momentum about any other axis (normal to z or not) cannot be conserved, for the potential energy varies with the degree of rotation. After all, the fact that we can refer to a rotation about some axis normal to the intermolecular vector as 'end-over-end' immediately implies that an 'end' has some meaning and—via the form of the potential energy at least—is different from the 'middle'. See the close of §7.3 for further remarks upon the non-conservation of angular momentum with respect to an axis that is not associated with a so-called infinite-fold rotation symmetry. Our conclusion is, therefore, that angular momentum is conserved *only* with respect to the internuclear molecular axis in the linear system; and that, while our choice of the label z is free, the physically unique axis is not. As the total orbital angular momentum is determined from the vectorial sum of, say, the three cartesian components and none but that about z is conserved, this total quantity, is therefore not conserved in linear molecules: $\mathcal{H}$ and l^2 do not commute. So while, via (2.15, 2.16) etc., l^2 and l_z share a common set of eigenfunctions, that set is not the one shared by $\mathcal{H}$ and l_z and so the quantum number l, and the fact that the m_l ranges l to $-l$ by integers, both lose relevance in the cylindrical molecular symmetry. The *only* orbital angular momentum that is of interest to us—the one conserved—is that with respect to the intermolecular axis. It is quantized, of course, and takes values 0, ±1, ±2, ±3 . . ., the $\pm$ signs corresponding to the two (and clearly only two) senses of rotation about z.

Obviously the extent of what can be said of angular momentum in even linear molecules is much less than that in atoms. There are few parallels. Nevertheless, the convention has arisen of labelling the angular momentum values above in a manner which appears to make an analogy with the scheme we have studied for atoms. While we must follow convention, it is important to be clear about the danger of confusion from the outset. The established nomenclature is to label orbitals in linear molecules—and the scheme is entirely restricted to systems with cylindrical symmetry—according to the moduli of their eigenvalues under l_z:

$$l_z\phi = \lambda\hbar\phi; \text{ linear systems} \tag{7.1}$$

$\lvert\lambda\rvert$	$= 0, 1, 2, 3 \ldots$
label	$\sigma, \pi, \delta, \phi \ldots$

The labels are, of course, the Greek equivalent to the Roman letters *s, p, d, f*. . . But while the latter label the positive values of l with respect to $\hat{l}^2$ in atoms, these Greek labels refer to the eigenfunctions of $\hat{l}_z$. There is more of a physical parallel between λ and m_l than between λ and l. So beware! Actually a much closer analogy is to be seen in the Stark splitting of atomic eigenfunctions in an electric field, as described in §4.11. There, the electric field was applied externally in the laboratory: here it arises internally through the presence of the neighbouring atom in the molecule.

7.2 *Spectroscopic versus chemical labels*

These same Greek labels will, of course, be familiar already through descriptions of bonding in organic and inorganic molecules. Their use is not quite the same, however, for the general chemist uses the nomenclature in a looser, less rigorous way than the spectroscopist or theoretician. Let us review the conventional, 'chemical' definitions.

Introductory definitions usually associate with these labels a varying degree of nodality as viewed down the internuclear axis: and discussions are usually centred upon the LCAO method for forming molecular orbitals. Thus, in Figure 7.1, we observe how the overlap of s and/or p_z orbitals on the centres of a diatomic molecule give rise to bonding or antibonding molecular orbitals possessing no nodal planes when viewed along the molecular symmetry axis. These functions, viewed along z, are circularly symmetric and called σ orbitals. The label refers to the molecular axis and so serves equally well for both the molecular orbitals and atomic orbitals in that environment. In Figure 7.2 are drawn some examples of π orbitals, so-called within the 'chemical' definitions because of the single nodal plane observed on viewing them along the internuclear vector. The first two MOs arise from overlap of, say, two p_x orbitals on adjacent centres in the usual bonding and anti-bonding manner. The last two sketches illustrate examples of bonding π overlap of d–p and d–d, respectively. Because the feature of sign-inversion of the orbital upon reflection in the plane containing the internuclear axis, and shown in the figure, is again common to both molecular orbitals and the constituent atomic orbitals, these various interactions shown in the figure can be labelled $p\pi$–$p\pi$, $d\pi$–$p\pi$ or $d\pi$–$d\pi$, as appropriate. Notice that identical diagrams can be drawn in the yz-plane also, implying a two-fold degeneracy of all π orbitals: we return to this issue shortly. Two nodal planes are observed for δ interaction, as shown in Figure 7.3. The diagram might represent the bonding overlap of two d_{xy} orbitals. If so, the same diagram, rotated by 45° around z, would serve for two $d_{x^2-y^2}$ orbitals. Note again that the equivalence of

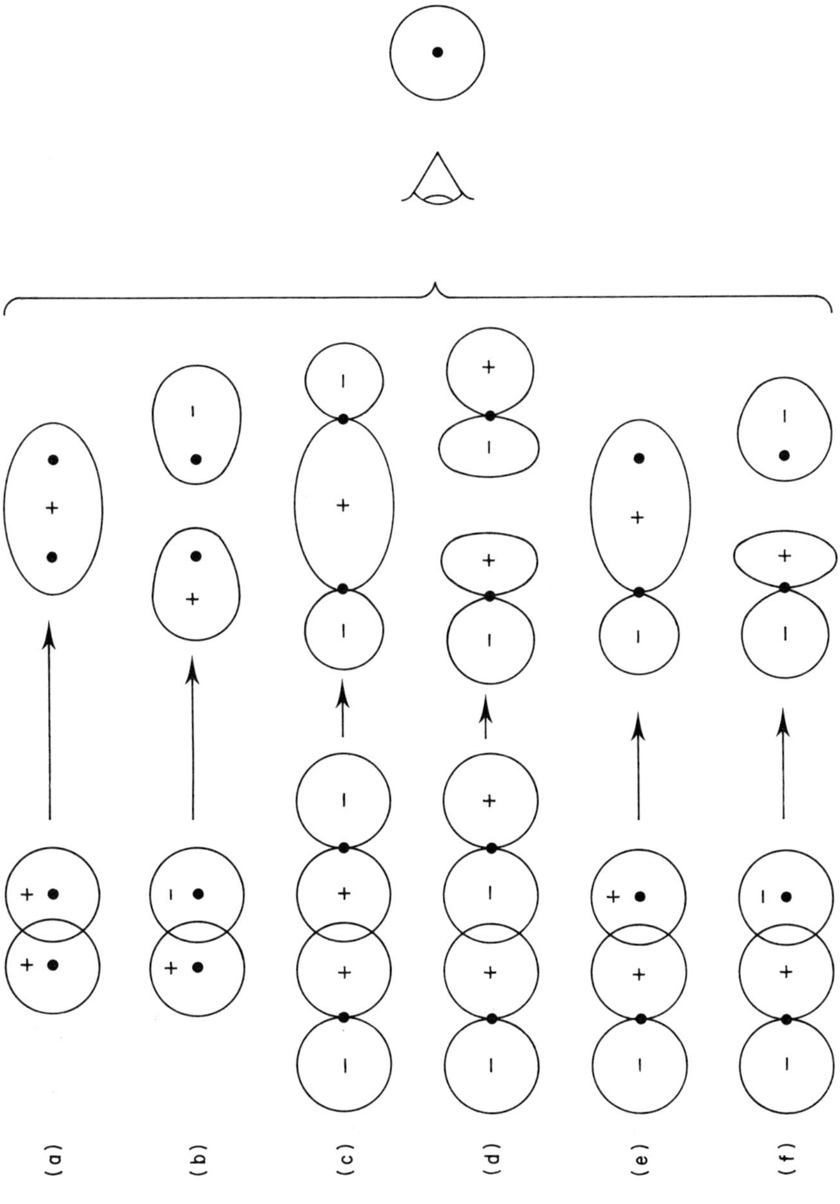

Figure 7.1 Showing various σ interactions. When viewed down the internuclear axis, these molecular orbitals are circularly symmetric.

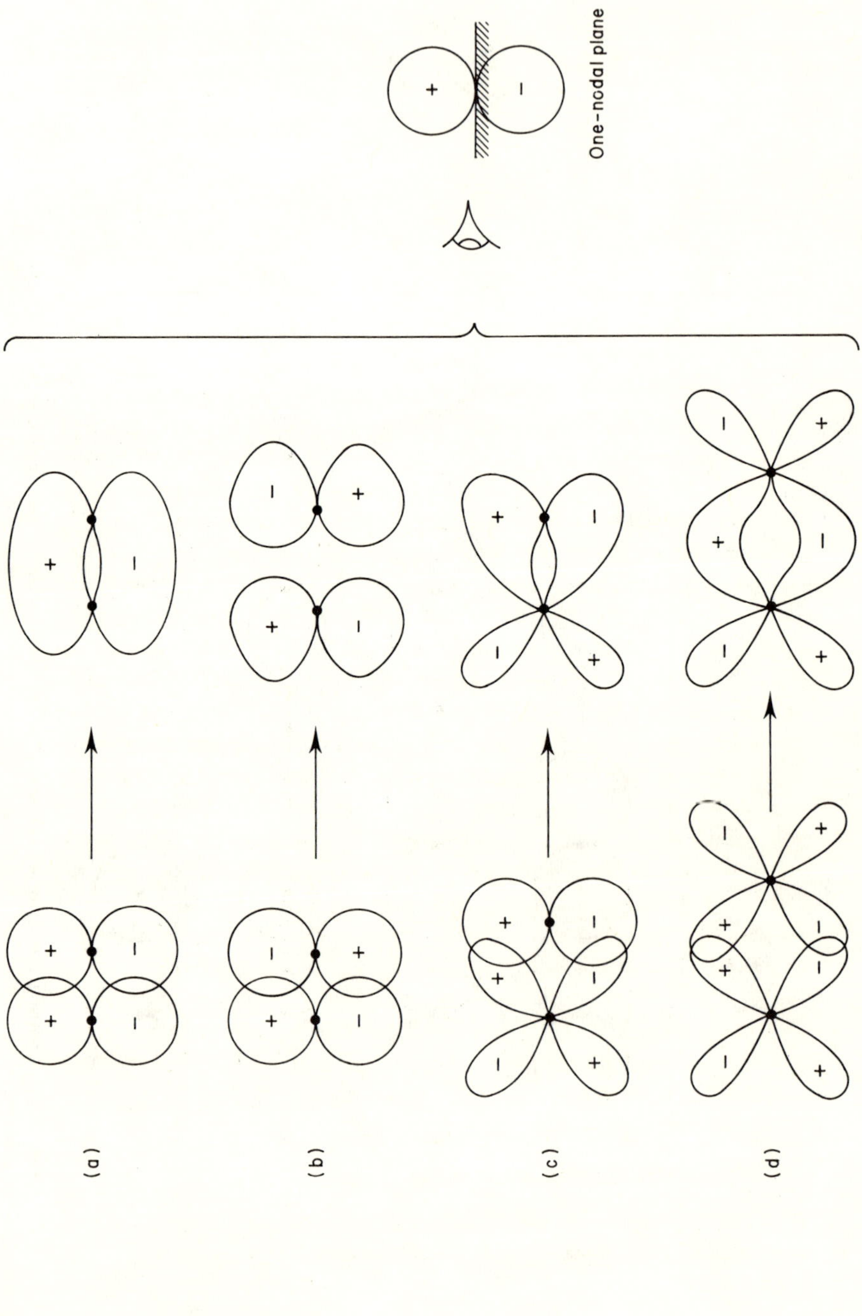

Figure 7.2 Various π interactions. Viewed down the internuclear axis, the molecular orbital exhibits one nodal plane.

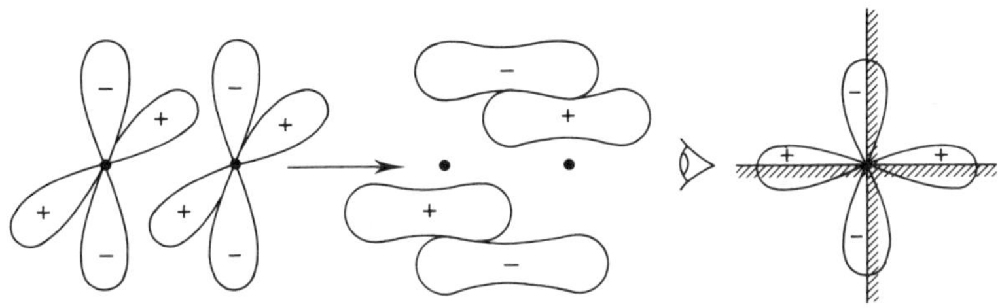

Figure 7.3 A δ bonding interaction. Viewed down the internuclear axis, the molecular orbit exhibits two nodal planes.

these two MOs implies a twofold degeneracy for δ-type interactions.

Notwithstanding different expositions, the chemical and spectroscopic definitions seem to be very much in accord, as they should be. Indeed, they would be, even if not for the common extension in mainstream chemistry discussions of using the same labels for molecules which are not cylindrically symmetric. A single example makes the point. The organic chemist describes the C–C bonding in acetylene in terms of one σ and two π bonds, while the spectroscopist would talk of one of each; similarly, that in ethylene is referred to by one σ and one π bond, which labels would be deemed incorrect spectroscopically within a non-axially symmetrical molecule. No conceptual difficulty arises in the common nomenclature, the labels σ and π essentially denoting little more than 'end-on' or 'sideways' overlap of the constituent atomic functions within the appropriate molecular orbitals. The idea is so well established that we must live with it. We need be rather more attentive, however, within the theoretical context, for there is no question but that one view is correct and technically 'natural' while the other is incorrect but seemingly easier on first acquaintance.

We are brushing alongside an all-pervasive feature of chemical theory, here, to do with the role of molecular symmetry and it is time we made the concept more explicit. So we will leave the question of the different definitions of π bonds in abeyance, returning to it in §7.5.

7.3 *Symmetry classification*

An object is symmetrical if it takes on an equivalent appearance as seen from different viewpoints. The idea can be formalized.

Consider, for example, the rotation of a plain square of card by 90° about a line passing through the middle of the card and at right angles to it: the result is an appearance which is indistinguishable from the original. The line about which the rotation was made exemplifies a *symmetry element*—here a fourfold rotation axis in a self-evident nomenclature—while the rotation itself is called a *symmetry operation*. A symmetry operation can, but need not, have several distinct parts. It can even have no parts at all as in the null or *identity* operation that leaves the object as it was: after all, if you closed your eyes while someone else performed the operation, you would be unable afterwards to distinguish an equivalent but different appearance of the object from an equivalent and identical one. All objects obviously possess the identity element but the relevance of every other symmetry element depends upon the system at hand. The plain square of card would be said to possess fourfold symmetry with respect to the normal to the centre of the card. It also possesses twofold symmetry with respect to that same axis; and also with respect to axes joining opposite corners of the card; and to those in the plane of the card, rotated by 45° with respect to these. There are many other symmetry elements characterizing such an object, the whole collection of elements forming a so-called *symmetry group*. We are at the foothills now of a subject called *Group Theory*, applications of which are very powerful in chemistry, but we do not describe it here, partly because it is so well-presented elsewhere[9] and partly because the few important features we now require can be introduced without a general structure.

The operation upon a system by some symmetry operator R yields an indistinguishable copy. Note, however, that both copy and original are *observable* entities. Therefore, while electron density or energy will be indistinguishable after a symmetry operation, wavefunctions need not be, for they are not themselves observable. So, in contemplating the operation

$$R\psi = \psi', \tag{7.2}$$

we enquire how ψ and ψ' may differ while their associated observables, energy and local electron density, say

$$E \quad = \langle\psi|\, \mathcal{H}\, |\psi\rangle \big/ \langle\psi/\psi\rangle \tag{7.3}$$

and

$$\rho \quad = \psi^*\psi, \tag{7.4}$$

are unchanged. It is clear that the symmetry operation R must change ψ by some constant c at most, so that $\psi' = c\psi$. It is equally obvious that the constant must have unit modulus. If c is real, it can take the values ± 1: but it can be more general. Recalling Table 6.1 and the associated discussion, we note that the equivalent of a real scalar in Euclidean space is a complex number in Hilbert space. The complex phase factor $e^{i\alpha}$, where α is some real number, has unit modulus and is the Hilbert space equivalent of the real phase factor ± 1: so we write

$$R\,\psi = e^{i\alpha}\psi. \tag{7.5}$$

The normalization of $\psi' = e^{i\alpha}\psi$ is unchanged:

$$\langle\psi'|\psi'\rangle = \langle e^{i\alpha}\psi|e^{i\alpha}\psi\rangle = e^{-i\alpha}e^{i\alpha}\langle\psi|\psi\rangle = 1, \tag{7.6}$$

assuming ψ itself to be normalized. The associated local electron density

$$\rho' = \psi'^*\psi' = e^{-i\alpha}\psi^*e^{i\alpha}\psi = \psi^*\psi = \rho, \tag{7.7}$$

is also unaltered by the phase factor, as is the energy:

$$\begin{aligned} E' \quad &= \langle\psi'|\,\mathcal{H}\,|\psi'\rangle \Big/ \langle\psi'|\psi'\rangle \\ &= e^{-i\alpha}e^{i\alpha}\langle\psi|\,\mathcal{H}\,|\psi\rangle \Big/ 1 = E, \end{aligned} \tag{7.8}$$

for α is not a parameter of $\mathcal{H}$. This latter point ensures that $e^{i\alpha}\psi$ is also an eigenfunction of $\mathcal{H}$ as the phase factor simply takes on the role of multiplier. It then follows that

$$\begin{aligned} \mathcal{H}\psi' &= \mathcal{H}(R\psi) = \mathcal{H}e^{i\alpha}\psi = e^{i\alpha}\mathcal{H}\psi = e^{i\alpha}E\psi \\ &= Ee^{i\alpha}\psi = E\psi' = ER\psi = RE\psi = R\mathcal{H}\psi, \end{aligned} \tag{7.9}$$

or, from the second and last terms, that

$$\mathcal{H}R\psi = R\mathcal{H}\psi. \tag{7.10}$$

This expression provides an alternative and more powerful *definition* of a symmetry operation; namely, as one which commutes with the Hamiltonian. The idea was anticipated in §2.8.

There is a very close parallel between the operators of angular momentum and so-called infinitesimal rotations. While we do not have the space here to develop this matter it is worth while remarking that both sets of operators share the same commutation relationships (2.12). Fleshing out our discussions in §7.1, we note that infinitesimal rotations—which produce an equivalent object upon endlessly small rotations about an infinite-fold rotation axis—are symmetry operators for all directions in spherical atoms. These rotation operators therefore commute with the atomic Hamiltonian: as noted in §7.1, so does l_z for any z direction. On the other hand, linear molecules possess only one such infinite-fold rotation axis and so it follows that only infinitesimal rotations about the internuclear vector commute with the cylindrical Hamiltonian. The same is true for the angular momentum operator. By the way, the fact that homonuclear diatomic molecules possess twofold rotation symmetry axes normal to the intermolecular vector is irrelevant in the present context, for the parallel between angular momentum and rotations is confined strictly to the case of *infinitesimal* rotations.

7.4 *Eigenvalues of symmetry operators*

Equation (7.5) expresses the fact that the symmetry operation R multiplies the wavefunction ψ by a number—$e^{i\alpha}$: it is an eigenvalue equation. However, while the eigenvalue equations we use for energy or angular momenta, for example, involve Hermitian operators which lead only to real eigenvalues—as required for observable quantities—the symmetry eigenvalues are not observable and may be complex. We need consider only two symmetry operators, here, and for these the eigenvalues happen to be real. For heuristic reasons, relating to remarks made near the end of §7.2, we discuss the application of these two operators to diatomic systems in a cyclic manner.

First, inversion in a centre. Homonuclear diatomic molecules are centrosymmetric; that is to say that they possess a symmetry element called a *centre of symmetry* or *inversion centre*, which makes equivalent, pairs of locations related diametrically through the molecular mid-point. As shown in Figure 7.4, this inversion symmetry reverses, or inverts, all coordinates. We use the symbol $\hat{\imath}$

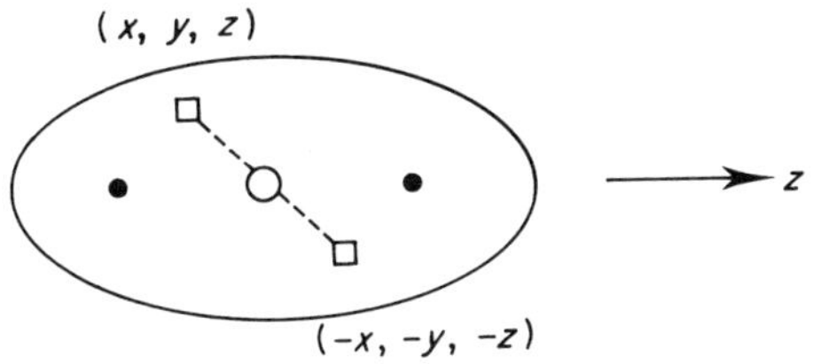

Figure 7.4 Inversion symmetry reverses the signs of all coordinates.

to represent the inversion operator, retaining the caret at all times to avoid confusion with the pure imaginary *i*. An eigenfunction of the homonuclear diatomic system may be unchanged or multiplied by −1 by the inversion operation:

$$\hat{\imath}\psi_g = +1\psi_g \tag{7.11}$$

$$\hat{\imath}\psi_u = -1\psi_u \tag{7.12}$$

and the symbols *g* and *u*—from the German *gerade* for even, or *ungerade* for odd—label the eigenvalues of the inversion operator. They may be regarded as symmetry quantum numbers. The molecular orbitals sketched in Figure 7.1(a) and (c) are examples of gerade orbitals; here labelled σ_g; while those in Figure 7.1(b), (d) are ungerade, σ_u. Those in Figure 7.1(e) and (f) are not related by an inversion centre and could not be stationary states in *homo*nuclear molecules. Note that for σ functions, the bonding molecular orbitals are gerade while the antibonding ones are ungerade. The reverse may be the case for the π functions shown in Figure 7.2. That in Figure 7.2(a) is bonding and labelled π_u while that in Figure 7.2(b) is π_g antibonding. The molecular orbital in Figure 7.2(c) is not a stationary state of a homonuclear system. A final remark at this point is to note how very simple and obvious are the assignments of the *g* and *u* labels to the functions shown in Figures 7.1 to 7.3.

Secondly we consider the case of reflection in a plane containing the internuclear axis. As all linear molecules, centrosymmetric or not, possess cylindrical symmetry, we recognize infinitely many planes containing the interatomic vector(s) with respect to which the molecule reflects into itself as exemplified in Figure 7.5.

It is to be emphasized, however, that the *xy* plane sketched in the figure is only one of the infinite number of equivalent, so-called, *mirror planes*. Convention has it that we call 'vertical' that direction

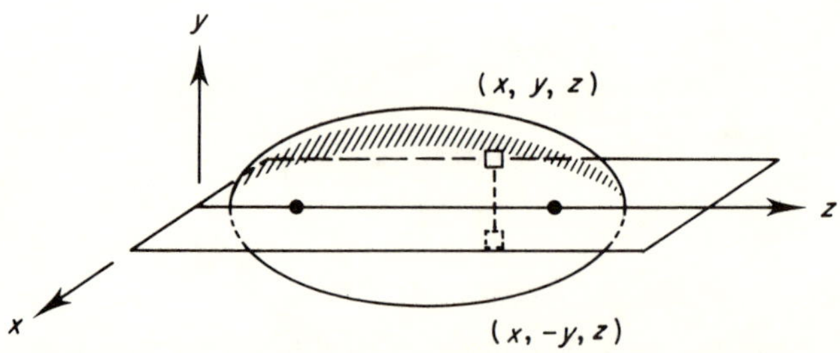

Figure 7.5 Reflection in a mirror plane reverses the sign of only one coordinate.

parallel to the rotation axis of highest order. In linear molecules this is the infinite-fold rotation axis along z, whether or not there is two-fold rotational symmetry about any normal to z, as in homonuclear diatomics. So we label the mirror plane as 'vertical', however it is drawn. A final complication in nomenclature now arises in that the accepted symbol for a mirror plane is σ! In order to minimize the possibility of confusion here, we shall only symbolize the *operation* of reflection in such planes—by the operator $\hat{\sigma}_v$ so that, once again, the caret will remove ambiguity. We read $\hat{\sigma}_v$ as the 'sigma-vee' operator and eigenfunctions of linear molecules may be unchanged or multiplied by -1 by it:

$$\hat{\sigma}_v\psi^+ = +1\psi^+ \tag{7.13}$$
$$\hat{\sigma}_v\psi^- = -1\psi^- \tag{7.14}$$

This time the label (or, in effect, symmetry quantum number) of the reflection eigenvalue—serving to label the associated eigenfunction—is a right-superscript of + or −, as shown.

Application of the idea to each of the σ orbitals (it is irrelevant whether they are g or u, of course) in Figure 7.1, shows them all to be even under $\hat{\sigma}_v$: they could be variously labelled σ_g^+, σ_u^+. What about the π orbitals in Figure 7.2? Well, inspection of the arrangement in Figure 7.6(a), illustrating the π_x molecular orbital as viewed along the internuclear axis (as in Figure 7.2 on the right) suggests that we should label it as π_x^- for the function changes sign on reflection in the mirror plane xz. On the other hand, if we choose the mirror plane yz—which is equivalent, remember—the function is unchanged and should be labelled π_x^+. the opposite conclusions would emerge with respect to the π_y orbital. We can formalize the results a little. For the choice of reflection in the xz plane, we have:

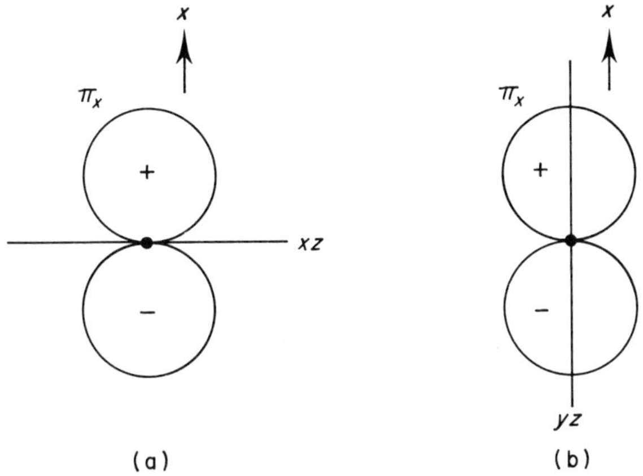

Figure 7.6 The mirrors xz and yz lie normal to the paper.

$$\text{xz mirror: } \hat{\sigma}_v \begin{pmatrix} \pi_x \\ \pi_y \end{pmatrix} \rightarrow \begin{pmatrix} -\pi_x \\ +\pi_y \end{pmatrix} \tag{7.15}$$

and for reflection in the yz plane, we find:

$$\text{yz mirror : } \hat{\sigma}_v \begin{pmatrix} \pi_x \\ \pi_y \end{pmatrix} \rightarrow \begin{pmatrix} +\pi_x \\ -\pi_y \end{pmatrix}. \tag{7.16}$$

But we need not stop there. Reflection in planes lying exactly between xz and yz, interchanges the π_x and, π_y orbitals, while multiplying them by +1 or −1 depending upon the choice of mirror, as shown in Figure 7.7.

Thus, for reflection in plane (3) of Figure 7.7(a):

$$\text{plane (3) : } \hat{\sigma}_v \begin{pmatrix} \pi_x \\ \pi_y \end{pmatrix} \rightarrow \begin{pmatrix} \pi_y \\ \pi_x \end{pmatrix}. \tag{7.17}$$

Similarly

$$\text{plane (4) : } \hat{\sigma}_v \begin{pmatrix} \pi_x \\ \pi_y \end{pmatrix} \rightarrow \begin{pmatrix} -\pi_y \\ -\pi_x \end{pmatrix}. \tag{7.18}$$

Equations (7.15) to (7.18) can be written as matrix equations in which the operator is *represented* by a different matrix on each occasion. Thus:

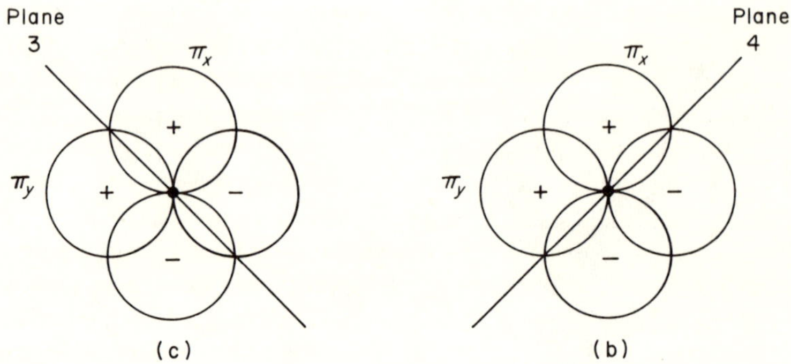

Figure 7.7 These mirrors interchange π_x and π_y: plane 4 also reverses their phases.

$$\begin{pmatrix} -1 & 0 \\ 0 & 1 \end{pmatrix}\begin{pmatrix} \pi_x \\ \pi_y \end{pmatrix} = \begin{pmatrix} -\pi_x \\ \pi_y \end{pmatrix}; \quad \begin{pmatrix} 1 & 0 \\ 0 & -1 \end{pmatrix}\begin{pmatrix} \pi_x \\ \pi_y \end{pmatrix} = \begin{pmatrix} \pi_x \\ -\pi_y \end{pmatrix} \tag{7.19}$$

$$\begin{pmatrix} 0 & 1 \\ 1 & 0 \end{pmatrix}\begin{pmatrix} \pi_x \\ \pi_y \end{pmatrix} = \begin{pmatrix} \pi_y \\ \pi_x \end{pmatrix}; \quad \begin{pmatrix} 0 & -1 \\ -1 & 0 \end{pmatrix}\begin{pmatrix} \pi_x \\ \pi_y \end{pmatrix} = \begin{pmatrix} -\pi_y \\ -\pi_x \end{pmatrix}$$

respectively. Recalling that we can choose any of the infinitely many equivalent reflection planes, it is obvious that in all but these most simple choices, the result of reflection is to scramble the π_x and π_y orbitals. For example, choosing on reflection plane 60° from xz could be represented by the matrix equation

$$\begin{pmatrix} \frac{1}{2} & \frac{\sqrt{3}}{2} \\ \frac{\sqrt{3}}{2} & -\frac{1}{2} \end{pmatrix}\begin{pmatrix} \pi_x \\ \pi_y \end{pmatrix} = \begin{pmatrix} \frac{1}{2}\pi_x + \frac{\sqrt{3}}{2}\pi_y \\ \frac{\sqrt{3}}{2}\pi_x - \frac{1}{2}\pi_y \end{pmatrix} \tag{7.20}$$

Problem
Verify (7.20)

We shall return to these matrix equations in a moment. For most purposes it is sufficient to recognize that we apparently get infinitely many different answers to our enquiry about the eigenvalues of π orbitals with respect to $\hat{\sigma}_v$. Indeed, in all cases other than (7.15) and (7.16), π_x and π_y were not even eigenfunctions for they changed to something quite different during the reflection operation. What have we done wrong? Which reflection plane should we take?

Well we began with the molecular orbitals shown in Figure 7.2. They are well known, easy, commonplace; but, in a somewhat technical sense, they are *not natural*. On the one hand, we insisted throughout that the cylindrical symmetry of linear molecules made all vertical mirror planes equivalent: on the other, we implied knowledge of, or ascribed significance to, the x and y directions by drawing the orbitals in Figures 7.6 and 7.7 and then labelling them. The π_x and π_y functions do not exemplify the axial molecular symmetry in a natural way. Instead we should work with orbitals π_{+1} and π_{-1}, the subscripts labelling their angular momentum quantum numbers (λ) with respect to z. Let us examine the relationships between π_{+1}, π_{-1}, π_x and π_y.

7.5 *Real and complex orbitals*

Using (2.21) and Table 2.1, we have the form of the angular parts of an atomic p orbital set.

$$p_0 \sim \cos\theta \qquad (7.21)$$

$$p_1 \sim -\sin\theta.\ e^{i\phi} \qquad (7.22)$$

$$p_{-1} \sim \sin\theta.\ e^{-i\phi} \qquad (7.23)$$

The $\sim$ sign indicates the neglect of normalization constants in the present discussion. The negative sign in front of the expression for p_1 arises from a *phase convention* due to Condon and Shortley such that we take this sign for those spherical harmonics with positive, odd m_l values. In *this* book, there is no need to follow the convention, provided we are consistent but in other applications, problems can arise so we may as well stick to it from the beginning.

As we saw in §2.6, p_0 looks like p_z and, since we are focusing upon the angular parts of these functions only, we shall so identify these functions:

$$p_z \equiv p_0. \qquad (7.24)$$

If we take the quantization axis to be the principal axis of a diatomic molecule also, we can equally label p_0 as p_σ, as in Figure 7.1(c) to (f). This follows because the angular momentum of p_0 about z is zero:

$$l_z p_0 = 0\hbar p_0 \tag{7.25}$$

The angular momenta of $p_{\pm 1}$ about z are just ± 1 (in atomic units, where $\hbar = 1$), which is why they are so labelled. However, let us pedantically work through this assertion. Dropping $\hbar$ from now on, we have from (2.38)

$$l_z|m_l\rangle = m_l|m_l\rangle. \tag{2.38}$$

Suppose we construct the matrix of p_1 and p_{-1} under l_z:

l_z	p_1	p_{-1}
p_1	$\langle p_1\|l_z\|p_1\rangle$	$\langle p_1\|l_z\|p_{-1}\rangle$
p_{-1}	$\langle p_{-1}\|l_z\|p_1\rangle$	$\langle p_{-1}\|l_z\|p_{-1}\rangle$

(7.26)

Each element of this matrix may be evaluated by operating on the ket using (2.38) and then multiplying by the bra $\langle m'_l|$. For example:

$$\langle p_1|l_z|p_1\rangle = \langle p_1|[1|p_1\rangle] = 1\,\langle p_1|p_1\rangle = 1, \tag{7.27}$$

assuming p_1 (and p_{-1}) are normalized. On the other hand,

$$\langle p_1|l_z|p_{-1}\rangle = \langle p_1|[-1|p_{-1}\rangle] = -1\langle p_1|p_{-1}\rangle = 0, \tag{7.28}$$

because p_1 and p_{-1} are orthogonal. Altogether, the table (7.26) takes the form:

l_z	p_1	p_{-1}
p_1	1	0
p_{-1}	0	-1

(7.29)

The matrix is diagonal and real with matrix elements describing the values of $+1$ and -1 unit of angular momentum about z, respectively, so justifying the labels $p_{\pm 1}$. But what do these orbitals look like?

From (7.22) and (7.23), we see that they are complex and so we cannot really draw them. Nevertheless, we will try as far as possible. First consider the Θ functions—here $\sin\theta$. In Figure 7.8(a) is traced the polar form of $\sin\theta$ as θ ranges from 0 to π. As the associated Legendre polynomial is independent of ϕ, the Θ function may be represented as a figure of revolution about the z-axis, that is everywhere positive for p_{-1} and everywhere negative for p_1 (because of the Condon–Shortley factor). While Figure 7.8(b) is as far as we can go with a conventional drawing, we can at least discuss the multiplicative Φ factors in (7.22) and (7.23). The Φ

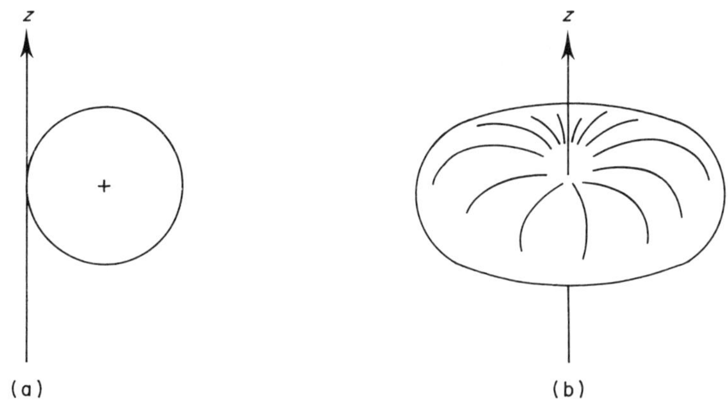

Figure 7.8 (a) The polar form of $\sin\theta$ for $\phi = 0$; and (b) the figure of revolution for $\sin\theta$ independent of ϕ.

functions take the form of phase factors of—in this case—unit modulus. For p_1; as ϕ varies from 0 to $\pi/2$ to π to $3\pi/2$, the phase factor changes from 1 to $+ i \sin\phi$ to -1 to $-i\sin\phi$. For p_{-1}, as ϕ covers the very same values, the phase factor changes from 1 to $-i\sin\phi$ to -1 to $+ i \sin\phi$. The phase changes are sketched in Figure 7.9 to emphasize the point that the phases change in opposite senses, corresponding to the different signs of the associated angular momenta.

So, while the angular momentum matrix (7.29) takes a simple and transparently obvious form, we have no particularly simple way of drawing the complex basis functions p_1 and p_{-1}. These two functions are, however, degenerate. We know that because all three p functions are degenerate in the free atom and the formation

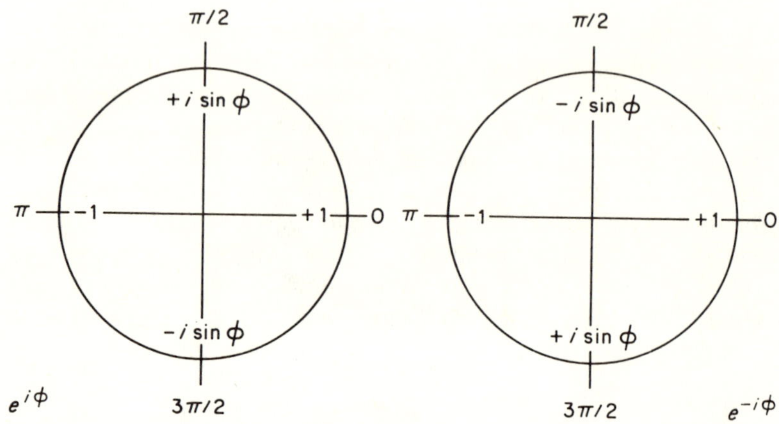

Figure 7.9 The phase factors $e^{\pm i\phi}$ shown as functions of the azimuthal angle ϕ.

of a linear molecule specializes only one axis. We now employ a simple theorem about degeneracy.

Theorem If two (or more) eigenfunctions of any operator are degenerate, so also is any completely arbitrary linear combination of them and this is also an eigenfunction.

Proof Let $\hat{O}\psi_1 = a\psi_1$ and $\hat{O}\psi_2 = a\psi_2$ describe the degenerate eigenfunctions ψ_1 and ψ_2 of the operator $\hat{O}$. Let $c_1\psi_1 + c_2\psi_2$, where c_1 and c_2 are any real or complex numbers, represent an arbitrary linear combination. Then

$$\begin{aligned} \hat{O}(c_1\psi_1 + c_2\psi_2) &= c_1 a\psi_1 + c_2 a\psi_2 \\ &= a(c_1\psi_1 + c_2\psi_2). \end{aligned} \qquad \text{Q.E.D.}$$

Comment The result does not mean that all degeneracies are infinite-fold degeneracies because only *n orthogonal* linear combinations can be constructed from *n* original functions. The choice of the first one is quite arbitrary, however. The totally arbitrary choice of a pair of degenerate functions (in the case of a twofold degeneracy) in Hilbert space is equivalent to the free choice of the *pair* of base vectors **i** and **j** in a two-dimensional Euclidean space.

Given, then, that p_1 and p_{-1} are degenerate we are at liberty to take any linear combinations of them we wish, knowing that they will remain eigenfunctions of the molecular Hamiltonian. So we combine them in such a way that we replace the complex functions of (7.22) and (7.23) with real ones. Suppose we add p_1 and p_{-1}, and renormalize to get p_a:

$$p_a = \frac{1}{\sqrt{2}}(p_1 + p_{-1}) \sim \sin\theta\,(-e^{i\phi} + e^{-i\phi})$$
$$\sim \sin\theta.\ -2i\sin\phi. \qquad (7.30)$$

This is not real, but nor is it complex: it can be made real on multiplication by i:

$$ip_a = \frac{i}{\sqrt{2}}(p_{-1} + p_1) \sim 2\sin\theta\sin\phi. \qquad (7.31)$$

We have established that sinθ looks like the doughnut in Figure 7.8(b). The polar form of sin ϕ is just the figure in the xy plane shown in Figure 7.10(a) by analogy with arguments used for Figure 2.5. Multiplication of sinθ by sinϕ then yields the form shown in Figure 7.10(b). It has *exactly* the same shape and parity characteristics of p_z but oriented along the y-axis: so we label it p_y.

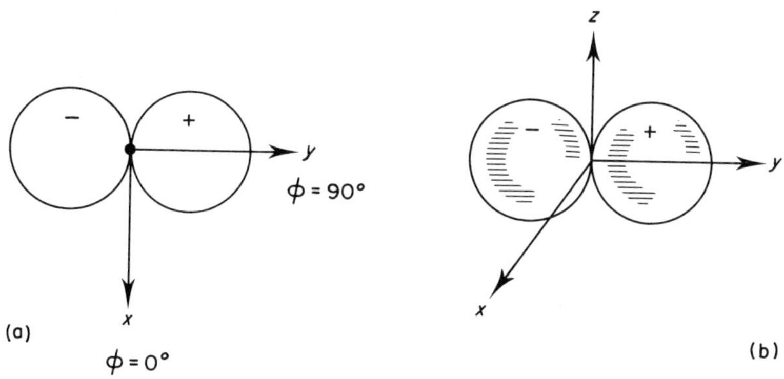

Figure 7.10 (a) The polar form of sin ϕ. (b) Multiplied by sin θ of Figure 7.8(b) to give p_y.

Similar arguments yield p_x as the real orbital of form sinθcosϕ given by the difference $(p_{-1} - p_1)$. In summary, we have

$$p_x = \frac{1}{\sqrt{2}}(p_{-1} - p_1) \qquad (7.32)$$

$$p_y = \frac{i}{\sqrt{2}}(p_{-1} + p_1), \tag{7.33}$$

as the real, drawable, p orbitals first encountered in elementary bonding discussions. Notice how they arise as a *pair* from two combinations of the *pair* of complex orbitals p_1 and p_{-1}.

Exercise Prove that p_x and p_y are orthogonal.

Now we construct a matrix equivalent to (7.26) but within the basis of *real* orbitals:

l_z	p_x	p_y
p_x	$\langle p_x\|l_z\|p_x\rangle$	$\langle p_x\|l_z\|p_y\rangle$
p_y	$\langle p_y\|l_z\|p_x\rangle$	$\langle p_y\|l_x\|p_y\rangle$

(7.34)

Consider the diagonal matrix element $\langle p_x|l_z|p_x\rangle$. We evaluate this after substitution of the expressions (7.32) and (7.33):

$$\begin{aligned}\langle p_x|l_z|p_x\rangle &= \langle\frac{1}{\sqrt{2}}(p_{-1}-p_1)|l_z|\frac{1}{\sqrt{2}}(p_{-1}-p_1)\rangle \\ &= \frac{1}{2}[\langle p_{-1}|\,l_z\,|p_{-1}\rangle - \langle p_{-1}|\,l_z\,|p_1\rangle - \langle p_1|\,l_z\,|p_{-1}\rangle \\ &\quad + \langle p_1|\,l_z\,|p_1\rangle] \\ &= \frac{1}{2}[-1-0-0+1] \\ &= 0,\end{aligned} \tag{7.35}$$

where we have used the information from (7.29) in the penultimate step. Now consider the off-diagonal matrix element $\langle p_x|l_z|p_y\rangle$:

$$
\begin{aligned}
\langle p_x|l_z|p_y\rangle &= \langle\frac{1}{\sqrt{2}}(p_{-1}-p_1)|l_z|\frac{i}{\sqrt{2}}(p_{-1}+p_1)\rangle \\
&= \frac{i}{2}[\langle p_{-1}|\,|p_{-1}\rangle + \langle p_{-1}|\,|p_1\rangle - \langle p_1|\,|p_{-1}\rangle \\
&\quad - \langle p_1|\,|p_1\rangle] \\
&= \frac{i}{2}[-1+0-0-1] \\
&= -i
\end{aligned}
\tag{7.36}
$$

The whole matrix must be Hermitian, for l_z is a Hermitian operator, whence:

l_z	p_x	p_y
p_x	0	$-i$
p_y	i	0

(7.37)

Exercise Demonstrate that $\langle p_y|l_z|p_x\rangle = +i$, *explicitly*.

Now when we compare the matrices (7.29) and (7.37) we observe different results. Within the complex basis $p_{\pm 1}$, the orbitals were not readily drawable but the angular momentum eigenvalues were real and obvious. In the present real basis of $p_{x,y}$, the orbitals are easy to visualize but the angular momentum—which must be the same because the orbitals, taken as a pair, are the same—is disguised in a pure imaginary and off-diagonal form. Of course (7.37) is just a more general form of (7.29) which we can show by diagonalization, as in §6.9. We write the secular determinantal equation as,

$$
\begin{vmatrix} 0-\lambda & -i \\ +i & 0-\lambda \end{vmatrix} = 0 \tag{7.38}
$$

and hence find,

$$(0 - \lambda)^2 - (-i)(+i) = 0$$

and

$$\lambda = \pm 1, \tag{7.39}$$

as in (7.29).

In concluding this section, therefore, we learn two points: First that, provided we take the p orbitals *as a pair*, they possess ± 1 unit of angular momentum about z whether they are expressed in the complex or real forms; and this rejustifies the label π corresponding to $\lambda = \pm 1$. Secondly, that the complex representation can be considered the more 'natural' for it immediately reveals the cylindrical symmetry of the axial molecule. We might say that the $p_{\pm 1}$ orbitals are *symmetry adapted*. The real forms might be easier to comprehend at first, and are certainly better known, but they lead, as we have seen, to a confusion with respect to the labelling under $\hat{\sigma}_v$.

7.6 *Review of orbital symmetry labels*

As the foregoing discussion of angular momentum referred to the molecular z-axis, all remarks concerning p orbitals carry over directly to those labelled π, for the spatial extension parallel to z corresponding to the formation of molecular orbitals alters none of it. Thus π_1 and π_{-1} are related to π_x and π_y by expressions exactly analogous to those in (7.32) and (7.33). Now we examine the result of operating upon $\pi_{\pm 1}$ with $\hat{\sigma}_v$ as promised at the end of §7.4. Referring to Figure 7.9, it is clear that the effect of *any* vertical mirror upon $e^{i\phi}$ is to transform it into $e^{-i\phi}$ and vice versa, because the operation reverses the sign of ϕ. This doesn't quite mean that $\pi_{\pm 1}$ are interchanged, however, because of the Condon–Shortley phase convention. Thus, by analogy with (7.22) and (7.23), we have

$$\hat{\sigma}_v \begin{pmatrix} \pi_1 \\ \pi_{-1} \end{pmatrix} \rightarrow \begin{pmatrix} -\pi_{-1} \\ -\pi_{+1} \end{pmatrix}. \tag{7.40}$$

which, written as a matrix equation, takes the form:

$$\begin{pmatrix} 0 & -1 \\ -1 & 0 \end{pmatrix}\begin{pmatrix} \pi_1 \\ \pi_{-1} \end{pmatrix} = \begin{pmatrix} -\pi_{-1} \\ -\pi_1 \end{pmatrix}. \tag{7.41}$$

As $e^{\pm i\phi}$ are interchanged by the mirror, the complex orbitals $\pi_{\pm1}$ are not individually eigenfunctions of $\hat{\sigma}_v$, so we do not attach a label like a $\pm$ superscript to them. It is not difficult to show that the same sort of result obtains for $\delta_{\pm1}$ orbitals and indeed for all the doubly degenerate species having $\lambda>0$.

Actually, we *could* label the pair of π orbitals with a zero right-superscript for the following reason. Inspection of *all* the various representations of the $\hat{\sigma}_v$ operator with respect to either real or complex basis functions—namely, (7.19), (7.20), (7.41)—reveals one common feature. All matrices have the same sum of leading diagonal elements (called the trace, Tr; spur, Sp; or character χ of a matrix): here, equal to zero. That sum is characteristic of the operation—hence the name 'character'. If we chose to regard the right superscript label of the molecular orbitals in linear molecules as the character of the matrix representing the σ_v operation in any given case, we would find a character of ±1 for the non-degenerate σ orbitals and of 0 for all twofold degenerate orbitals, π, δ . . . .

In practice, we do not use these labels at all for linear orbitals! This is because they are redundant in view of the previous summarizing sentence: a parallel example is the non-use of the doublet spin labels for orbitals—^2s, ^2p, etc., as discussed in §3.2. Nevertheless, none of the present discussions are wasted as we shall see when we consider molecular *terms* in the next chapter.

Finally, we have seen that the generality of (7.41) arises naturally when we apply a representative symmetry operation ($\hat{\sigma}_v$) to natural basis functions: the embarrassments arising from the use of the easy but unnatural real basis do not arise. For consistency's sake we should re-examine the situation with respect to the inversion operator within the complex π basis and show that all is well. In Figure 7.11 are sketched, as well as may be in view of the complex forms of $\pi_{\pm1}$, bonding and antibonding combinations of atomic π functions in a diatomic molecule. The bonding MO is characterized by in-phase ϕ functions, the antibonding one by out-of-phase ϕ terms. By reference to Figure 7.9, if the value of $e^{i\phi}$ at point (1) is $+i \sin \phi$, then that at (1′), related to (1) diametrically through the inversion centre C, is $-i \sin \phi$: there is a similar

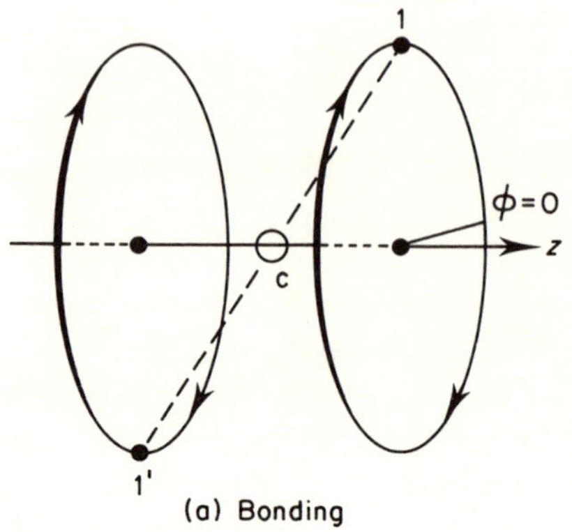

Figure 7.11 Inversion in a centre, C, for the complex forms $\pi_{\pm 1}$ gives the same result as for the real forms π_x and π_y in the bonding and antibonding arrangements of Figure 7.2(a) and (b).

inversion for all points so related and we therefore label the bonding π orbital once more as π_u. On the other hand, the points (2) and (2′) in Figure 7.11(b) have equal phases—on reference to Figure 7.9 again—and hence the antibonding π molecular orbital is labelled π_g. While we might just as well continue to determine the parity labels by the traditional route using *real* orbitals, as in §7.4, the present discussion serves to confirm that we do have a consistent and complete approach within the spectroscopically 'natural', complex basis.

CHAPTER 8

Diatomic Molecular States

8.1 *Diatomic molecular configurations*

The construction of molecular orbital diagrams, especially those for diatomic molecules is widely described and may be supposed well enough known to merit the briefest of reviews here. Consider the case of first-row homonuclear diatomic molecules, Li_2 to Ne_2. A first look at the combination of the atomic orbitals $1s$, $2s$, $2p$ on each atom suggests the molecular orbital sequence shown in Figure 8.1. Regarding the nomenclature of the MOs two small points can be made beyond the σ/π and g/u classification. It is common, though not mandatory, to number the orbitals in energy sequence from the lowest. For example, $3\sigma_g$ means the third highest σ_g molecular orbital in the stack: it does not need to arise from the third highest atomic orbitals, as here—see Figure 8.2. Secondly, the asterisk denotes the antibonding nature of the MO with respect to the parent AOs: it is not a symmetry label and, again, is not mandatory.

The relative orbital splittings and orderings given in the diagram follow from our first view using informed guesswork. Overlap between the tightly bound $1s$ orbitals is expected to be minimal and indeed the small splitting shown in the figure is included only for reasons of completeness: in many diagrams $1s$–$1s$ interaction is ignored and we focus on the the valence shells rather than on the cores. The overlap of $p\pi$ orbitals on the two centres is expected to be considerably less than between $p\sigma$, or even $2s$ functions so we draw the π_u–π_g separation as rather less than the various σ_g–σ_u splittings. With a little less conviction we estimate the splitting between $2s$–$2s$ overlap to be less than that between the more directed $2p_z$–$2p_z$ combination: on the other hand, only those halves of the p_z orbitals directed towards the other atom will be

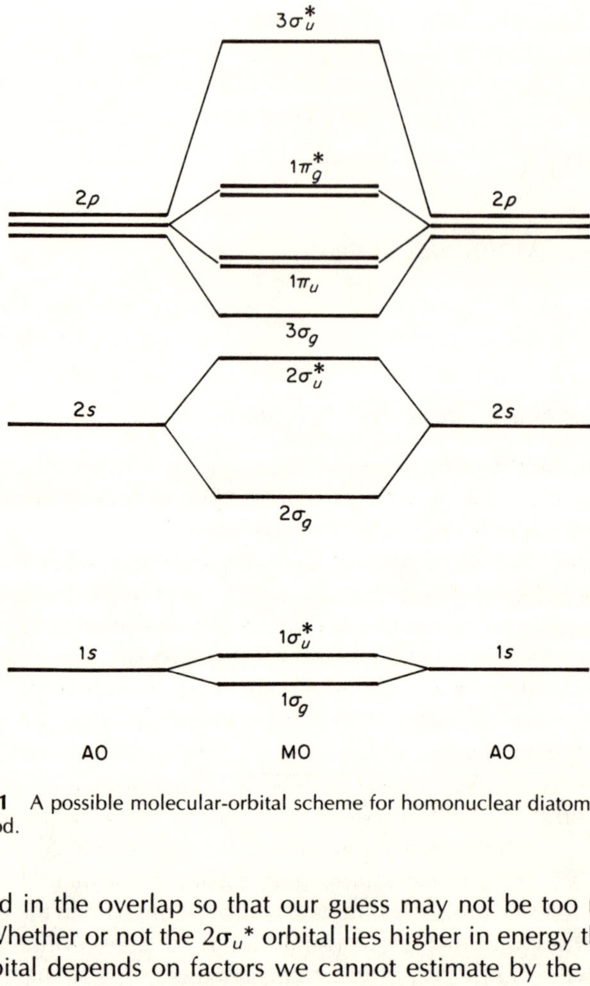

Figure 8.1 A possible molecular-orbital scheme for homonuclear diatomics of the first period.

involved in the overlap so that our guess may not be too reliable here. Whether or not the $2\sigma_u{}^*$ orbital lies higher in energy than the $3\sigma_g$ orbital depends on factors we cannot estimate by the present level of guesswork and, further, a more detailed study raises the possibility that the $3\sigma_g$ orbital may lie higher than the $1\pi_u$. Briefly, the reasoning goes as follows.

If the promotion energy $2s \rightarrow 2p$ is sufficiently large compared with the overlap—and hence resonance—integrals, we will find $2\sigma_u{}^*$ lying below $3\sigma_g$. However, as we traverse the period Li→Ne, the effective nuclear charge increases and this serves both to increase the $2s$–$2p$ promotion energy (because the more penetrating $2s$ orbital is then even more tightly bound relative to the $2p$) and to decrease the overall atomic radius. In turn the latter may or may not lead to better overlap so that without quantitative calculations or empirical, experimental evidence, we cannot be sure of the result of these conflicting trends. In addition to this consideration, we note that the $2\sigma_g$ and $3\sigma_g$ orbitals are of identical symmetries and, via perturbation theory say, may mix. In terms of the language of §6.5, these levels repel one another to an extent inversely proportional to their original energy separation. For several diatomic molecules of the first period, the orbital ordering is actually better represented as in Figure 8.2.

The $2\sigma_g$–$3\sigma_g$ 'repulsion' can alternatively be viewed as resulting from a hybridization of the atomic 2*s* and $2p_z$ on each centre followed by overlap of the resulting atomic *sp* hybrids. The choice of appropriate orbital sequences in any given case is best determined from experimental study of the resulting molecular terms: we return to this point in §8.6. A precursor to a description of such terms is the allocation of electrons to molecular orbitals that we call a configuration.

When discussing atoms, 'configuration' describes the many-electron wavefunctions that arise when we neglect interelectron correlation and spin-orbit coupling. In the same way, electron configurations in a molecule arise by assigning electrons to the

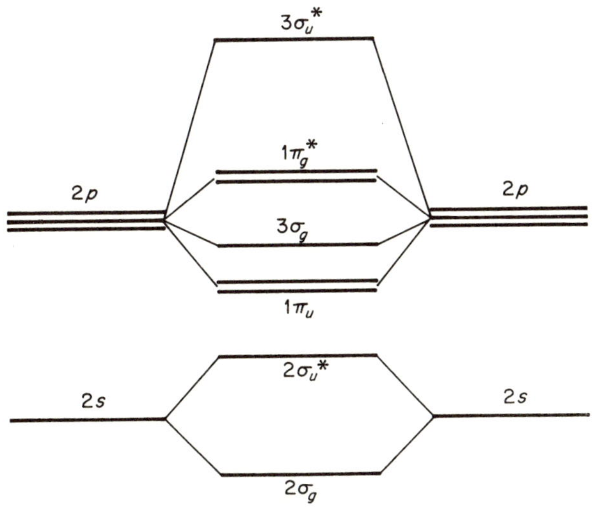

Figure 8.2 An alternative ordering of the valence MOs in first-period diatomics.

various molecular orbitals (for the ground configurations, using the Aufbau principle) taking no account of electron correlations. If the orbital sequence for O_2, for example, is as shown in Figure 8.2, the 16 electrons fill orbitals to yield the lowest energy configuration, $(1\sigma_g)^2$ $(1\sigma_u{}^*)^2$ $(2\sigma_g)^2$ $(2\sigma_u{}^*)^2$ $(1\pi_u)^4$ $(3\sigma_g)^2$ $(1\pi_g)^2$. The ground configuration of the $0_2{}^+$ ion would be written, $(1\sigma_g)^2$ $(1\sigma_u{}^*)^2$ $(2\sigma_g)^2$ $(2\sigma_u{}^*)^2$ $(1\pi_u)^4$ $(3\sigma_g{}^2)$ $(1\pi_g)^1$. Many excited configurations can be written down also: for example, for 0_2: $(1\sigma_g)^2$ $(1\sigma_u{}^*)^2$ $(2\sigma_g)^2$ $(2\sigma_u{}^*)^2$ $(1\pi_u)^3$ $(3\sigma_g)^2$ $(1\pi_g)^3$. Each of these configurations parallels the Ar$3d^2$ configuration of the V^{3+} ion, for example. In Chapter 3 we examined the terms arising from the d^2 atomic configuration. Here we now consider the terms arising from configurations in diatomic molecules.

8.2 *Closed shells and open shells*

Consider the ground configuration proposed above for the O_2 molecule: all orbitals except the last are completely filled. We refer to the partial configuration consisting of filled orbitals as involving *closed shells*, and incompletely filled orbitals (or combinations thereof) as *open shells*. All closed shells are characterized by the same 'symmetry' designations, as follows.

Just as the lower case letters σ, π, δ . . . refer to the orbital or one-electron angular momentum property about the internuclear vector as the only such conserved quantity, so we use the upper case Greek alphabet Σ, Π, Δ. . . to label the equivalent term-, or many-electron property. Again beware of the false analogy with the capital Roman letters *S, P, D*. . . for exactly the same reasons as stated at the end of §7.1. Now, as orbital angular momentum is conserved in linear molecules only with respect to the molecular axis, the total, many-electron quantity is given simply as the sum of contributions from the individual electrons in the configuration. For the closed shells of the O_2 molecule, we find a value (0×2) for $(1\sigma_g)^2$; plus (0×2) for $(1\sigma_u{}^*)^2$, plus $(0 \times 2) + (0 \times 2)$; plus $[2 \times (+1)] + [2 \times (-1)]$ for the $(\pi_{+1})^2$ and $(\pi_{-1})^2$ contributions; that is, zero. It is obvious that we always get zero: σ electrons make a zero contribution anyway, and π_{+1} and π_{-1} electrons make cancelling contributions provided they are equally occupied—and they must be in a closed shell. Altogether, therefore, Λ, the angular momentum about z for *any* closed shell configuration in a linear molecule, is zero, and we write the letter Σ for this.

Even more obvious is the fact that all electron spins are paired up in any closed shell, so that $\Sigma m_s = 0$ and $S = 0$. So we arrive at the description $^1\Sigma$.

Now consider the configuration symmetry under the inversion centre. Quite generally, if we multiply an even by an even function, the result is even: an odd by an odd is likewise even: but an odd times an even is odd. The rule,

$$g \times g = g; \quad u \times u = g; \quad u \times g = u \tag{8.1}$$

is a general mathematical property, not restricted to molecular orbitals. In *any* closed shell configuration (diatomic or not), the number of one-electron wavefunctions of odd parity must be even (the same is true for the even-parity functions but that fact is not required). Therefore, products of $u \times u$ occur in pairs and the total closed shell configuration parity is *even*. In centrosymmetric (homonuclear) diatomics, then, the ground configuration is characterized by the symbols, $^1\Sigma_g$.

Finally, we enquire about the right superscript, + or −. We again arrive at an answer by considering a product, as for the parity label. After all we are asking what happens as we operate upon the closed-shell configuration with the reflection, $\hat{\sigma}_v$. Suppose we had a configuration (a^+) (b^-)—which from our summary in §7.6 we know to be impossible actually—then the operator $\hat{\sigma}_v$ changes (a^+) into $+(a^+)$ and (b^-) into $-(b^-)$, for that is what the labels mean. Therefore $(a^+ \; b^-)$ goes to $-(a^+b^-)$: in short, we multiply the superscripts. All σ orbitals are even under the $\hat{\sigma}_v$ reflection so any configuration product of σ orbitals will be even also. Any filled π shell has an equal number of electrons occupying π_{+1} and π_{-1} orbitals. We know that $\hat{\sigma}_v$ interchanges these labels and so

$$\hat{\sigma}_v \; (\pi_{+1}\pi_{-1}) \rightarrow (\pi_{-1}\pi_{+1}) \tag{8.2}$$

and the reflection symmetry of the *product* is even. The same would be true of any pair of δ orbitals also. Our final characterization of *any* closed shell for a homonuclear, diatomic molecule is, therefore, $^1\Sigma_g{}^+$. For a heteronuclear system, which lacks the inversion centre, the label is $^1\Sigma^+$.

Returning to the example of O_2: we may divide the configuration and characterization symbols as follows:

$$\underbrace{(1\sigma_g)^2(1\sigma_u{}^*)^2(2\sigma_g)^2(2\sigma_u{}^*)^2(1\pi_u)^4(3\sigma_g)^2}_{\text{Closed shells}} \quad \underbrace{(1\pi_g)^2}_{\text{Open shell}} \qquad (8.3)$$

$$^1\Sigma_g{}^+ \qquad\qquad ^{2S+1}X$$

The characterization of the total configuration—closed times open—is just the same as that (those) for the open shell(s). This follows because:

(i) S for the closed shell is zero and so adds nothing to that for the open shell,
(ii) Λ is zero for the closed shell and adds nothing to that for the open one either,
(iii) The parity of the open shell will be unchanged on multiplication by the g of the closed shell,
(iv) The same is true for the symmetry of X with respect to $\hat{\sigma}_v$ on being multiplied by the even + superscript of the closed shell.

In summary: the spin and space labels of terms arising from any given configuration are identical to those determined for the open shells alone. The conclusion is quite generally true for atoms, and molecules of any shape. We employed the idea implicitly in Chapter 3 when considering the configuration of V^{3+} as $3d^2$ rather than Ar$3d^2$.

8.3 *The open shell* $(\pi_g)^2$

Two 'chemical' orbitals are represented by π_g but four spin-orbitals. Hence we consider the distribution of two electrons amongst four functions and that can be done in $^4C_2 = 4!/2!2! = 6$ ways. This is a perfect analogy for the $^{10}C_2 = 45$ ways of arranging two atomic d electrons. We sketch them in Figure 8.3, noting that we place both electrons, with their spins paired, in each of the $\pi_{g\pm1}$ orbitals; or one in each. Apart from the degeneracy of the orbitals, the situation is just the same as that shown in Figure 5.1 and we can carry over most of the subsequent discussion also but we must include now the question of the reflection $\hat{\sigma}_v$.

The orbital angular momentum, with respect to the diatomic axis, of the electronic wavefunction product π_{g+1} ($\uparrow\downarrow$)—that is, of

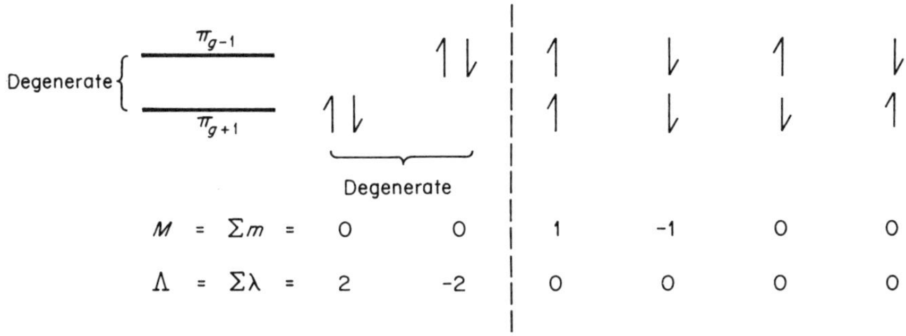

Figure 8.3 Compare the spin-singlets and triplets in Figure 5.1. The degeneracy of the first two singlets arises directly from the degeneracy of the $\pi_{g\pm1}$ orbitals.

$(\pi_{g+1}\alpha)$ $(\pi_{g+1}\beta)$—is given as usual by the sum of the individual electronic properties: here $+2$ units. Similarly that of $(\pi_{g-1}\alpha)$ $(\pi_{g-1}\beta)$ is -2 units. Operating with $\hat{\sigma}_v$ upon $(\pi_{g+1}\alpha)$ $(\pi_{g+1}\beta)$ gives

$$\hat{\sigma}_v\,[\pi_{g+1}\alpha\pi_{g+1}\beta] = [-\,\pi_{g-1}\alpha]\,[-\pi_{g-1}\beta]$$
$$= [\pi_{g-1}\alpha\pi_{g-1}\beta], \qquad (8.4)$$

using (7.40). In other words, reflection in a vertical mirror plane interchanges the first two electronic arrangements given in Figure 8.3 so that they transform *as a pair*:

$$\hat{\sigma}_v \begin{pmatrix} \pi_{g+1}\alpha\pi_{g+1}\beta \\ \pi_{g-1}\alpha\pi_{g-1}\beta \end{pmatrix} \rightarrow \begin{pmatrix} \pi_{g-1}\alpha\pi_{g-1}\beta \\ \pi_{g+1}\alpha\pi_{g+1}\beta \end{pmatrix}. \qquad (8.5)$$

And this accords well with their degeneracy, for this pair of electronic products, possessing plus and minus two units of orbital angular momentum, comprise a Δ term. The complete characterization for the term is $^1\Delta_g$ because (i) the spins are necessarily paired so that $S = 0$, (ii) two electrons in the same orbital must describe states of even parity (this does not depend upon the orbital here being gerade), and (iii) all terms other than Σ are twofold degenerate and transform as a pair, always in the manner of (8.5), so that we get a zero value for the character of the matrix representing $\hat{\sigma}_v$, as in §7.6, and we therefore attach no superscript label to the term symbol to record the fact. Finally, note that in order to make the product wavefunctions in $^1\Delta_g$ properly antisymmetric, they must be written

$$\begin{cases} [\pi_{g+1}(1)\pi_{g+1}(2)]\,\dfrac{1}{\sqrt{2}}\,[\alpha(1)\beta(2)-\alpha(2)\beta(1)] \\ \\ [\pi_{g-1}(1)\pi_{g-1}(2)]\,\dfrac{1}{\sqrt{2}}\,[\alpha(1)\beta(2)-\alpha(2)\beta(1)]. \end{cases} \tag{8.6}$$

We learn rather more from the last four products in Figure 8.3. We know from §5.2 that they span a spin-triplet and a spin-singlet. Here, each arrangement involves one electron in each of $\pi_{\pm 1}$ orbitals, so that $\Lambda = 0$ only: therefore, we know the terms arising from the $(\pi_{g+1})^1$ $(\pi_{g-1})^1$ configuration are partly described as $^1\Sigma_g$ and $^3\Sigma_g$—the gerade symbol follows very simply. Before moving on to discuss the operation of $\hat{\sigma}_v$, let us be clear on what we have deduced so far. The last few arrangements depicted in Figure 8.3 span the configuration $(\pi_{g+1})^1$ $(\pi_{g-1})^1$. Each possesses zero orbital angular momentum, corresponding to there being only one way of placing two electrons in two orbitals—ignoring spin; and only a Σ term is onefold degenerate. Focusing on the spin, we observe that there are, as in §5.2, four ways of arranging the electrons—ignoring space (not that that caveat makes any difference in the present case, but we shall see another kind of circumstance later); and these account for the *spin* singlet and triplet—one plus three.

By analogy with §5.2, the spin-singlet state wavefunction is written,

$$\psi_S{}^0 = \frac{1}{\sqrt{2}}[\pi_{+1}(1)\pi_{-1}(2) + \pi_{+1}(2)\pi_{-1}(1)]\,\frac{1}{\sqrt{2}}\,[\alpha(1)\beta(2) - \alpha(2)\beta(1)] \tag{8.7}$$

where we have dropped the g subscripts for typographical convenience. Operating upon this function with the reflection $\hat{\sigma}_v$ interchanges π_{+1} and π_{-1} functions, as usual:

$$\hat{\sigma}_v\psi_S{}^0 = \frac{1}{\sqrt{2}}\,[\pi_{-1}(1)\pi_{+1}(2) + \pi_{-1}(2)\pi_{+1}(1)]\,\frac{1}{\sqrt{2}}\,[\alpha(1)\beta(2) - \alpha(2)\beta(1)]$$

$$= \psi_S{}^0. \tag{8.8}$$

(note that the two minus signs of (7.40) cancel). As the state $^1\Sigma_g$ thus reflects into itself times +1 we say that it is even under the $\hat{\sigma}_v$ reflection and label it finally as $^1\Sigma_g{}^+$.

On the other hand, a typical component of the spin-triplet term is written, by analogy with (5.13), as

$$\psi_T^1 = \frac{1}{\sqrt{2}}\,[\pi_{+1}(1)\pi_{-1}(2) - \pi_{+1}(2)\,\pi_{-1}(1)]\;\alpha(1)\alpha(2) \tag{8.9}$$

and the operation of $\hat{\sigma}_v$ upon this gives,

$$\hat{\sigma}_v\psi_T^1 = \frac{1}{\sqrt{2}}\,[\pi_{-1}(1)\pi_{+1}(2) - \pi_{-1}(2)\pi_{+1}(1)]\alpha(1)\alpha(2)$$

$$= -\psi_T^1. \tag{8.10}$$

Note, by the way, that $\hat{\sigma}_v$ is a space-only operator and does not affect the spin functions: therefore the behaviour of each component of the spin triplet is identical—the spin functions in (7.49) and (7.51) 'go along for the ride'. So we see that the spin-triplet functions are each transformed into their negatives: they are odd under the $\hat{\sigma}_v$ reflection and so are labelled $^3\Sigma_g{}^-$. This behaviour followed from the negative sign in the space part of the wavefunction in (8.9). Antisymmetry with respect to electron interchange here goes hand in hand with antisymmetry with respect to reflection in a vertical mirror but observe that in one case it is the electrons that are permuted while in the other, it is the orbitals.

Figure 8.4 summarizes the terms arising from the configuration $(\pi_g)^2$—and hence, by §7.8, from the ground configuration of the O_2 molecule: it may be compared with Figure 3.4 for the terms arising from the atomic d^2 configuration. Hund's rules place the $^3\Sigma_g{}^-$

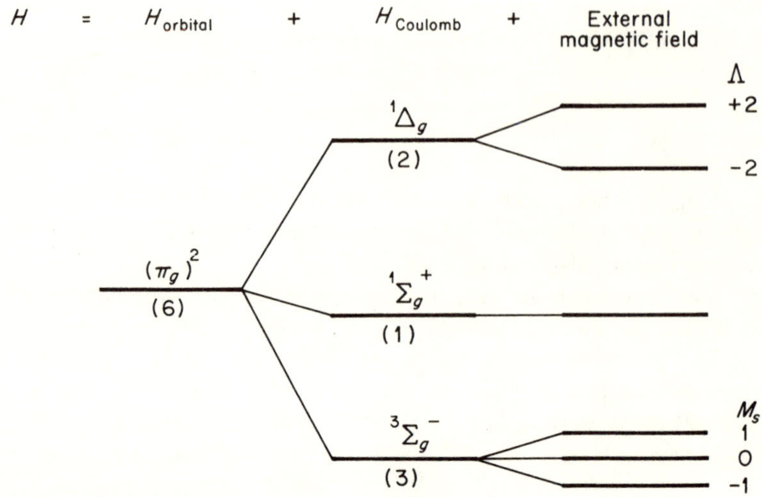

Figure 8.4 Spectroscopy in the presence of an external magnetic field could distinguish the terms arising from the $(\pi_g)^2$ configuration.

lowest in energy but do not assist in placing the remaining terms. As usual, absolute term energies would be available either from detailed and lengthy numerical computations or from appropriate experiments. In this connection, the most 'appropriate' measurement would be those of a qualitative nature which could identify the terms from their degeneracies.

In an externally applied magnetic field the terms split as shown in Figure 8.4. Zeeman spectroscopy would then show splittings of the normal $^3\Sigma_g^- \rightarrow {}^1\Sigma_g^+$, $^1\Delta_g$ spectral transitions, corresponding to excitations from each component of the ground term to each component of the various excited ones. To repeat, however, neither absolute nor relative energies of any excited term are accessible to the angular momentum theory we have discussed in this book.

8.4 *The open shell* $(\sigma_u)^1(\pi_g)^3$

The three terms just discussed label six different electronic arrangements arising from double occupancy of the four spin-orbitals spanning the configuration $(\pi_g)^2$. Their energies differ by virtue of the Coulomb interelectron part of the total molecular Hamiltonian. They are not the only electronic arrangements open to the oxygen molecule, however, any more than $Ar3d^2$ describes all possibilities for V^{3+} ions. Other configurations are possible also and each of those gives rise to one or more terms under the Coulomb contribution. For example, an excited configuration of O_2 arises if we promote an electron out of a (σ_u) orbital into the previously half-filled (π_g) set:

$$(1\sigma_g)^2(1\sigma_u^*)^2(2\sigma_g)^2(2\sigma_u^*)^1(1\pi_g)^4(3\sigma_g)^2(1\pi_g)^3. \qquad (8.11)$$

Using the arguments of §8.2, we determine the terms arising from this configuration by study of the open shells, $(2\sigma_u^*)^1(1\pi_g)^3$, which by dropping unnecessary labels, we write now as $(\sigma_u)^1$ $(\pi_g)^3$. It clearly does not matter for us whether the σ_u orbital is higher or lower in energy than the π_g. In order to condense the process now, and also to set forth a 'recipe' for it, we proceed stepwise.

Degeneracy and holes

The σ_u orbital comprises two spin-orbitals and the π_g, four: we have four electrons to distribute amongst them. However, we do *not* write the number of arrangements as ${}^6C_4$ for these would include those with, say, all four electrons in the π_g set and none in the σ_u: while those are possible, they do not belong to the configuration currently being studied. Instead, the total degeneracy of the $(\sigma_u)^1$ $(\pi_g)^3$ configuration is given by the product of the degeneracies for the separate configuration $(\sigma_u)^1$ and $(\pi_g)^3$; that is, ${}^2C_1 \times {}^4C_3 = 2 \times 4 = 8$.

Next we might contemplate drawing electronic arrangements like those in Figure 8.3—though we could avoid that at the cost of a little more abstraction—but if so we shall end up placing three arrows in the pair of orbitals labelled $\pi_{\pm 1}$ (note that we shall drop and recover the g label as convenient, from now on). Some effort can be saved, and maybe mistakes avoided, if we make use of the hole formalism here. This would allow us to replace $(\pi_g)^3$ by $(\pi_g)^1$ for the number and detail of the arrangements of three electrons within the four spin-orbitals of π_g are precisely the same as those for one hole in a $(\pi_g)^4$ shell. The detailed argument simply parallels that in §4.5. In particular, the degeneracy of $(\pi_g)^3$ is ${}^4C_3 = 4!/3!1!$ while that of $(\pi_g)^1$ is ${}^4C_1 = 4!/1!3!$.

In summary, the first step is to recognize the open shell; to consider again the hole equivalent of any part of it that involves less particles; and finally to calculate the total degeneracy against which the final answers may be checked later.

Space and spin

The eight-particle arrangements in $(\sigma_u)^1\ (\pi_g)^1$ could be drawn out explicitly as in Figure 8.5. While it is perfectly possible to deduce what terms are spanned by these arrangements as they stand, we can simplify matters considerably by taking advantage of the separation of space and spin functions. That separation is possible under the neglect of spin-orbit coupling we have made in this chapter so far. We make brief mention of spin-orbit coupling in diatomic molecules at the end, meanwhile noting that the neglect is justified in practice because most work on diatomics, and certainly for us, concerns the first row (Li–Ne) for which spin-orbit coupling coefficients are very small: hence splittings due to the effect are usually negligible even when they are not exactly zero.

Saying that space and spin functions are separable means that the total wavefunction may be factorized into space and spin parts, as in (8.7) for example. In turn we may consider the electronic arrangements with respect to space and spin quite separately from one another. Consider first the space part. Here we draw the diagram in Figure 8.6, in which we represent electrons or holes by dots rather than arrows so as to ignore their spin property. The arrangements shown in Figure 8.6 are the only ones possible—spacewise—for the $(\sigma)^1\ (\pi)^1$ configuration and obviously require the label Π. Recalling that one electron resides in a σ_u orbital and one

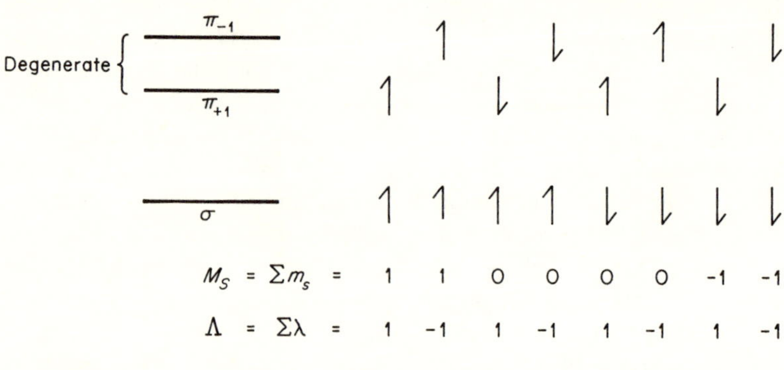

Figure 8.5 The eight-particle arrangements in the configuration $(\sigma)^1(\pi)^1$.

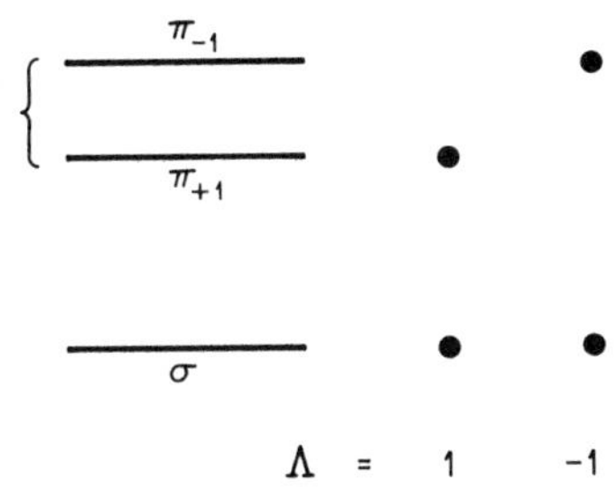

Figure 8.6 The space part of the particle distribution in $(\sigma)^1(\pi)^1$.

hole in a π_g, we can immediately add the parity description to get Π_u. Having dealt with the space part, we consider the spin using Figure 8.7. Here we do not bother to recognize the two parts of the degenerate π shell for that is irrelevant (and an unnecessary complication) in describing the spin possibilities. The pattern in Figure 8.7 is just like that on the right of Figures 8.3 or 5.1. The pattern occurs so often in this area of our subject and this is one reason for introducing its analysis in §5.2 in so general a way: here we have made the equivalence with the discussion in §5.2 by setting $a \equiv \sigma$, $b \equiv \pi$. Hence we know that the four spin arrangements give rise to one spin-singlet and one spin-triplet. As this observation is quite separate from that which led to the label Π_g above, the space description applies equally to each spin label—or vice versa. Therefore, we find $(\sigma_u)^1\ (\pi_g)^1$ gives rise to $^1\Pi_u + {}^3\Pi_u$. As usual, there are no superscript labels.

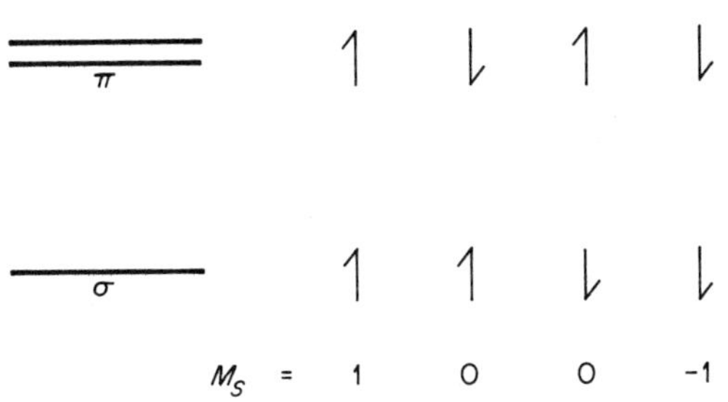

Figure 8.7 The spin part of the particle distribution in $(\sigma)^1(\pi)^1$.

Check degeneracy

The degeneracies of these terms are 1 × 2 and 3 × 2, respectively giving a total of 8 as required. Hund's rule places ${}^3\Pi_u$ lower than ${}^1\Pi_u$ but the energies of neither term relative to any but the ground term of the ground configuration $(\pi_g)^2$ are determined by such means. In summary, we have Figure 8.8.

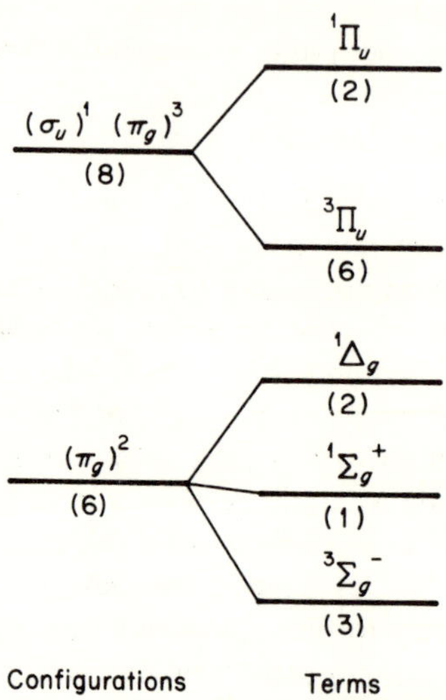

Figure 8.8 Hund's rules determine the lowest-lying term from each configuration.

8.5 *The open shell* $(\pi_u)^3\ (\pi_g)^3$

Various such exercises are to be found in various other books. They can all be solved using the procedures described in the preceding section. One further complication arises, however, when we consider the configuration $(\pi_u)^3\ (\pi_g)^3$ which can be another excited configuration of the O_2 molecule. For all practical purposes at this level, a mastery of this example tests the student's understanding of several of the topics we have discussed and should leave him feeling that all other (and more usual) examples are simple by comparison.

Degeneracy and holes

The hole equivalent of $(\pi_u)^3$ $(\pi_g)^3$ is $(\pi_u)^1$ $(\pi_g)^1$ and has fewer particles to consider. The total degeneracy is ${}^4C_1 \times {}^4C_1 = 16$.

Space and spin

Spatial arrangements are shown in Figure 8.9 and can be accounted for by one Δ term plus two Σ: for the moment, let us label these Σ (i) and Σ (ii).

The spin arrangements are just the same as in Figures 8.7 or 5.1, reading in the latter case, $a \equiv \pi_g$ and $b \equiv \pi_u$. So we have one singlet and one triplet. As space and spin parts are independent, we have a spin triplet and singlet for each of Δ, Σ (i) and Σ (ii). The parity of all terms is u, arising as they do from one u hole and one g

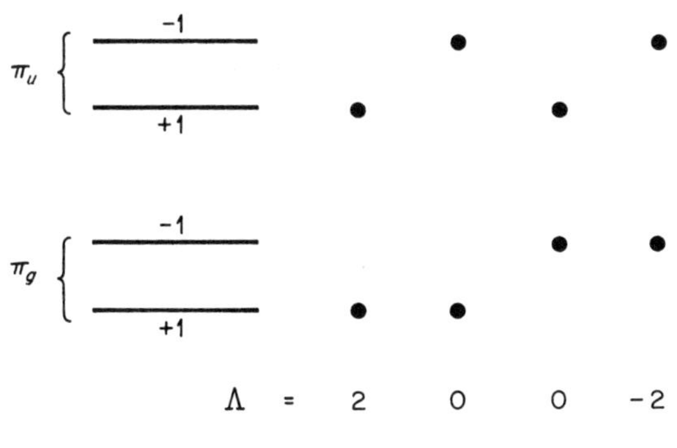

Figure 8.9 Spatial arrangements in the configuration $(\pi_u)^1(\pi_g)^1$.

hole. The Δ terms have no right superscript as may be checked once more by direct operation of $\hat{\sigma}_v$ which will be found to interchange the ± 2 components.

There remains the problem of characterizing the Σ terms with respect to $\hat{\sigma}_v$. Consider the $M_S = 1$ component of the ${}^3\Sigma$ (i) term:

$$\psi^1[{}^3\Sigma(\text{i})] = \frac{1}{\sqrt{2}}\,[\pi_{g+1}(1)\pi_{u-1}(2) - \pi_{g+1}(2)\pi_{u-1}(1)]\,\alpha(1)\alpha(2). \tag{8.12}$$

Operation with $\hat{\sigma}_v$ gives;

$$\hat{\sigma}_v \{\psi^1[^3\Sigma(\mathrm{i})]\} = \frac{1}{\sqrt{2}}[\pi_{g-1}(1)\pi_{u+1}(2) - \pi_{g-1}(2)\pi_{u+1}(1)]\,\alpha(1)\alpha(2)$$

$$= \psi^1\ [^3\Sigma(\mathrm{ii})]. \tag{8.13}$$

It is easy to show that the operation of $\hat{\sigma}_v$ upon Σ (ii) describes the inverse. So far we have expected reflection in a vertical mirror to convert a Σ term into itself or minus itself: and indeed this must still be so. As derived so far, however, we find $\hat{\sigma}_v$ interchanges the two $^3\Sigma_u$ terms, and the same is true, by exact analogy, of the two $^1\Sigma_u$ states. It is no accident though, that the two (triplet or singlet) Σ states are degenerate—that derives from the degeneracy of the π orbitals and the nature of the configuration—so that we may combine them arbitrarily, as discussed in §7.5. In particular, we want to form linear combinations that *are* symmetric and antisymmetric with respect to $\hat{\sigma}_v$. We can guess what combinations to try. Let $^3\Sigma_u$(a) label the combination,

$$^3\Sigma_u(\mathrm{a}) = \frac{1}{\sqrt{2}}\ [^3\Sigma_u(\mathrm{i}) + {}^3\Sigma_u(\mathrm{ii})] \tag{8.14}$$

and $^3\Sigma_u$(b), its orthogonal counterpart:

$$^3\Sigma_u(\mathrm{b}) = \frac{1}{\sqrt{2}}\ [^3\Sigma_u(\mathrm{i}) - {}^3\Sigma_u(\mathrm{ii})]. \tag{8.15}$$

We examine their behaviour under reflection symmetry:

$$\hat{\sigma}_v\ [^3\Sigma_u(\mathrm{a})] = \frac{1}{\sqrt{2}}\ [^3\Sigma_u(\mathrm{ii}) + {}^3\Sigma_u(\mathrm{i})], \tag{8.16a}$$

from (7.54) and so,

$$\hat{\sigma}_v[^3\Sigma_u(\mathrm{a})] = +\,1\ [^3\Sigma_u(\mathrm{a})] \tag{8.16b}$$

and we relabel ${}^3\Sigma_u$(a) as ${}^3\Sigma_u{}^+$. Similarly,

$$\hat{\sigma}_u\ [{}^3\Sigma_u(\text{b})] = \frac{1}{\sqrt{2}}\ [{}^3\Sigma_u(\text{ii}) - {}^3\Sigma_u(\text{i})]$$

$$= -1\ [{}^3\Sigma_u(\text{b})] \tag{8.17}$$

and so ${}^3\Sigma_u(\text{b}) \equiv {}^3\Sigma_u{}^-$.

Check degeneracy

Collecting all our results together we find the situation in Figure 8.10, Hund's rule once more giving ${}^3\Delta_u$ as the ground term for this configuration. The total degeneracy of ${}^3\Delta_u + {}^1\Delta_u + {}^3\Sigma_u{}^+ + {}^3\Sigma_u{}^- + {}^1\Sigma_u{}^+ + {}^1\Sigma_u{}^-$ is 16, as it should be.

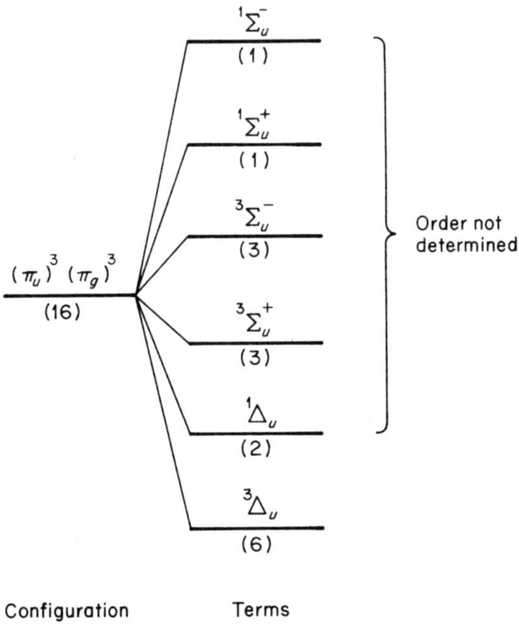

Figure 8.10 The complete set of terms arising from the configuration $(\pi_u)^3(\pi_g)^3$.

For completeness, to show the complex nature of the result we have derived with rather little effort, let us look at the form of the wavefunction for the $M_S = 0$ component of the ${}^3\Sigma_u{}^-$ term:

$$\frac{1}{2}\ [\pi_{g+1}(1)\pi_{u-1}(2)\ -\ \pi_{g+1}(2)\pi_{u-1}(1)\ -\ \pi_{g-1}(1)\pi_{u+1}(2)\ +$$
$$\pi_{g-1}(2)\pi_{u+1}(1)]\ \tfrac{1}{\sqrt{2}}\ [\alpha(1)\beta(2)\ -\ (\alpha(2)\beta(1)]$$

which, in an unfactorized form is,

$$\frac{1}{2\sqrt{2}}\ [\pi_{g+1}\alpha(1)\pi_{u-1}\beta(2) - \pi_{g+1}\beta(1)\pi_{u-1}\alpha(2) - \pi_{g+1}\beta(2)\pi_{u-1}\alpha(1)$$
$$+\ \pi_{g+1}\alpha(2)\pi_{u-1}\beta(1)\ -\ \pi_{g-1}\alpha(1)\pi_{u+1}\beta(2)\ +\ \pi_{g-1}\beta(1)\pi_{u+1}\alpha(2)$$
$$+\ \pi_{g-1}\beta(2)\pi_{u+1}\alpha(1)\ -\ \pi_{g+1}\alpha(2)\pi_{u+1}\beta(1)]; \qquad (8.19)$$

clearly a none-too-convenient expression!

8.6 *Diatomic molecular states—concluding remarks*

In the last few sections we have studied just three of the very many configurations of the O_2 molecule. The same or equivalent open shells occur for several other diatomic species also; and there are, of course, many others. The point here is to note the degrees of complexity (not difficulty) that arise from such simple beginnings. Somewhat greater complexity is shown by the spectra of atomic and molecular transition metal species but, for various reasons, this is little pursued: in complexes, for example, broad bands replace sharp spectral lines and much detail is lost. Even so, within the ground d^n configurations of transition metal complexes, many term splittings can be determined spectroscopically and their study forms part of the subject called ligand-field theory. That theory is really only applicable to states arising from the ground d^n configuration of those species and the inadequacy of similarly simple theories to account for states arising from other configurations in those systems is one reason why attention has been focused almost exclusively on the ground configuration.

By way of example, in Figure 8.11 are shown potential energy curves for *some* of the electronic states in C_2. The shapes, energy ordinates and relative displacements of those curves have all been established by painstaking studies of rotation, vibration and electronic spectroscopy by means which are no part of our brief in this book. Those curves constitute the sort of information ultimately

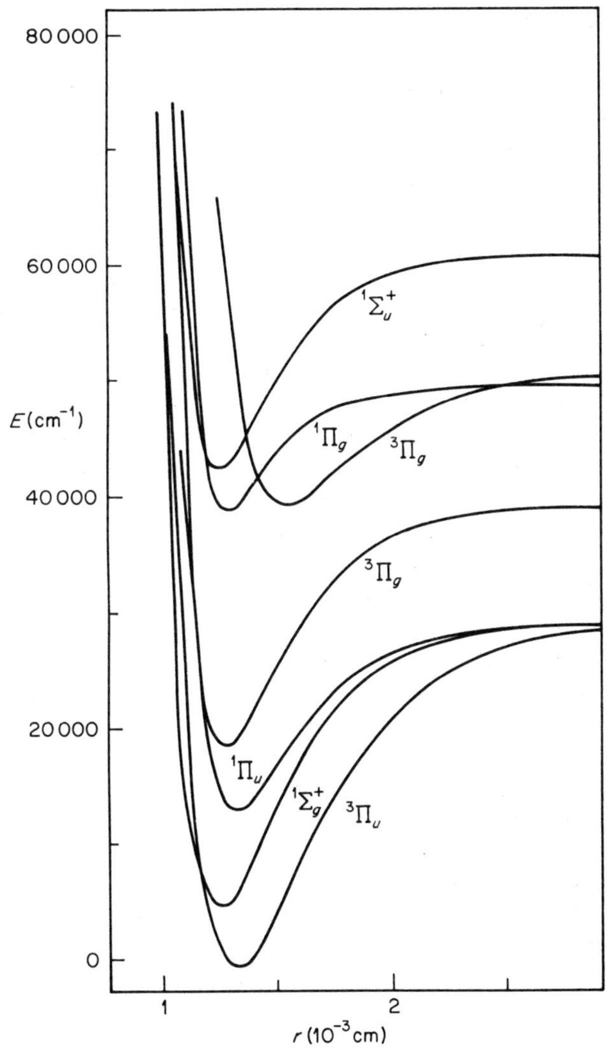

Figure 8.11 Potential energy curves for *some* of the electronic states in C_2.

available to us through spectroscopy. To some extent we can use these results to comment upon the ordering that must have been present in underlying molecular orbital diagrams like that in Figure 8.2. It is *not* an automatic process, however: it is not even a uniquely defined one. For while the observed properties of the molecular states are defined (how else are they observed?), no feature of the 'underlying' orbitals is observable. Properties of the orbitals in a one-electron system (also, effectively, in a one-electron open shell outside of cores) *can* be observed, but that is only because these orbitals are (trivially) the states in those systems, by identity. Otherwise, the orbitals are merely bases and, hence, arbitrary. *If* we propose that the molecular states that we observe are to be considered as composed of such-and-such explicitly defined molecualr orbitals, then comparisons of diagrams like that in Figure 8.11, but for the various first-row diatomic species, say, will lead to firm conclusions about the relative orderings of these molecular orbitals in each species. But the final significance of such conclusion depends, as they say, from where you start. All this is no council of despair for the molecular states *are* accessible to fundamental theory. Quantitative aspects (energies, dipole moments, and so on) of the molecular states that we have constructed from orbital occupancy are not determined by the qualitative procedures we have studied.

We have barely considered spin-orbit coupling in connection with diatomic systems. As mentioned earlier this is primarily because spin-orbit coefficients are very small in the early part of the periodic table. Another reason lies in the fact that the occurrence of either spin- or orbital-singlet terms in this area is very common. When either S or Λ, or both, are zero, the vector coupling rule leads to a unique coupled state; in short, no spin-orbit splitting. However, when neither S nor Λ are zero, coupling will take place but between Λ and the z-component of S and thus in a manner not exactly analogous to the L, S coupling described in Chapter 4. Although this will be important in diatomic transition metal species, for example, it is usually of little importance generally in chemistry, as noted above, and so we do not study it further here.

References and Further Reading

1. J. C. Polkinghorne, *The Quantum World*, Longman, London, 1984.
2. B. Hoffman, *The Strange Story of the Quantum*, Penguin, New York, 1947.
3. H. Eyring, J. Walter and G. E. Kimball, *Quantum Chemistry*, Wiley, New York, 1944.
4. P. W. Atkins, *Molecular Quantum Mechanics*, Oxford University Press, Oxford, 1970.
5. B. N. Figgis, *Introduction to Ligand Fields*, Wiley, New York, 1966.
6. C. J. Ballhausen, *Introduction to Ligand-field Theory*, McGraw-Hill, New York, 1962.
7. M. Gerloch, *Magnetism and Ligand-Field Analysis*, Cambridge University Press, London, 1983.
8. M. Gerloch and R. C. Slade *Ligand-Field Parameters*, Cambridge University Press, London, 1973.
9. F. A. Cotton, *Chemical Applications of Group Theory*, Wiley, New York, 1963.

INDEX